MATHEMATICS RESEARCH DEVELOPMENTS

NAVIER-STOKES EQUATIONS AND THEIR APPLICATIONS

MATHEMATICS RESEARCH DEVELOPMENTS

Additional books and e-books in this series can be found on Nova's website under the Series tab.

MATHEMATICS RESEARCH DEVELOPMENTS

NAVIER-STOKES EQUATIONS AND THEIR APPLICATIONS

PETER J. JOHNSON
EDITOR

Copyright © 2021 by Nova Science Publishers, Inc.
DOI: https://doi.org/10.52305/UJUZ9424

All rights reserved. No part of this book may be reproduced, stored in a retrieval system or transmitted in any form or by any means: electronic, electrostatic, magnetic, tape, mechanical photocopying, recording or otherwise without the written permission of the Publisher.

We have partnered with Copyright Clearance Center to make it easy for you to obtain permissions to reuse content from this publication. Simply navigate to this publication's page on Nova's website and locate the "Get Permission" button below the title description. This button is linked directly to the title's permission page on copyright.com. Alternatively, you can visit copyright.com and search by title, ISBN, or ISSN.

For further questions about using the service on copyright.com, please contact:
Copyright Clearance Center
Phone: +1-(978) 750-8400   Fax: +1-(978) 750-4470   E-mail: info@copyright.com

NOTICE TO THE READER

The Publisher has taken reasonable care in the preparation of this book, but makes no expressed or implied warranty of any kind and assumes no responsibility for any errors or omissions. No liability is assumed for incidental or consequential damages in connection with or arising out of information contained in this book. The Publisher shall not be liable for any special, consequential, or exemplary damages resulting, in whole or in part, from the readers' use of, or reliance upon, this material. Any parts of this book based on government reports are so indicated and copyright is claimed for those parts to the extent applicable to compilations of such works.

Independent verification should be sought for any data, advice or recommendations contained in this book. In addition, no responsibility is assumed by the Publisher for any injury and/or damage to persons or property arising from any methods, products, instructions, ideas or otherwise contained in this publication.

This publication is designed to provide accurate and authoritative information with regard to the subject matter covered herein. It is sold with the clear understanding that the Publisher is not engaged in rendering legal or any other professional services. If legal or any other expert assistance is required, the services of a competent person should be sought. FROM A DECLARATION OF PARTICIPANTS JOINTLY ADOPTED BY A COMMITTEE OF THE AMERICAN BAR ASSOCIATION AND A COMMITTEE OF PUBLISHERS.

Additional color graphics may be available in the e-book version of this book.

Library of Congress Cataloging-in-Publication Data

ISBN: 978-1-53619-967-3

Published by Nova Science Publishers, Inc. † New York

Contents

Preface		**vii**
Chapter 1	Kinetic Monism and All-Unity in Russian Cosmism versus Newtonian Dualism of Separated Energies *Igor Bulyzhenkov*	**1**
Chapter 2	Simulation of High-Temperature Flows in Nozzles with Unsteady Local Energy Supply *N. Brykov, V. Emelyanov and K. Volkov*	**25**
Chapter 3	Integrals of the Navier - Stokes and Euler Equations for Motion of Incompressible Medium *Alexander V. Koptev*	**61**
Chapter 4	Deep Water Movement *Alexander V. Koptev*	**83**
Index		**103**

PREFACE

In physics, Navier-Stokes equations are the partial differential equations that describe the motion of viscous fluid substances. In this book, these equations and their applications are described in detail. Chapter One analyzes the differences between kinetic monism and all-unity in Russian cosmism and Newtonian dualism of separated energies. Chapter Two presents a model for the numerical study of unsteady gas dynamic effects accompanying local heat release in the subsonic part of a nozzle for a given distribution of power of energy. Chapter Three describes a study of relationships between integrals and areas of their applicability. Lastly, Chapter Four defines the exact solutions of the Navier-Stokes equations characterizing movement in deep water and near the surface.

As explained in Chapter 1, the all-unity and kinetic monism of continuous matter in the Russian Cosmism contradicts conceptually the Newtonian dualism of kinetic and gravitational energies that ruined the intellectual legacy of many Russian thinkers. Lomonosov super-penetrating liquid, Umov inertial ether, Gurvish nonlocal biofields, Vernadsky Noosphere of thoughts, Chizhevsky cosmic pulsations, Bekhterev immortality of thoughts, Tsiolkovsky cosmic life and other orthodox suggestions are still incomprehensible and not in demand by the science textbooks and education curriculum based on negative (nonexistent) gravitational energies and the questionable "action-at-a-distance." Starting from the local cause of all accelerations according to Lomonosov and the hidden variable for extracting mass-energy from the inertial ether according to Umov, everyone can logically replace the fiction of gravitational potentials with non-local self-assembling of correlated densities in the volume integral of kinetic energy. Mathematical superposition of continuous material densities, inaccessible to human perception and measurements, can deduce the inverse square law for mutual accelerations and revive many "weird" ideas of Russian cosmists. The kinetic cooperation of correlated densities and stresses of continuous mass-energy with absolute self-knowledge of its non-local distribution is a more fundamental component of cosmic human beings than their visual structure with the densest volumes in spatial communications. Self-control of adaptive kinetic densities sets in motion the material all-unity of nonlocal bodies and invisible distributions of thoughts over the Noosphere. The kinetic monism of nonlocal energy can verifiably modify Euler / Navier-Stokes fluids by inertial feedback and remove non-Euclidean 3-interval, empty space, gravitational energy, black holes, and dark matter from educational programs in physics, astronomy, biology, medicine, and other natural sciences.

As shown in Chapter 2, a model for the numerical study of unsteady gas dynamic effects accompanying local heat release in the subsonic part of a nozzle for a given distribution of power of energy release is developed. The finite volume method is applied to solve unsteady compressible Navier-Stokes equations with high-temperature gas effects. Solutions of some benchmark test cases are reported, and comparison between computational results of chemically equilibrium and perfect air flowfields is performed. The constructed model is applied to flows with unsteady energy supply in nozzles, which are of interest for alternating current plasmatrons. Moving arcs are formed which are sources of intense energy supply. It is allowed to burn one or more arcs that are at different points in space, have different intensities and move at a given speed relative to the flow. A qualitative structure of shock wave and thermal processes in the nozzle during unsteady energy supply is discussed. The results of numerical simulation of one-dimensional and two-dimensional under- and overexpanded nozzle flows with a moving region of energy supply are presented. Output nozzle parameters are calculated as functions of a number and time of burning of plasmatron arcs. The results obtained show a qualitative pattern of gas dynamics and thermal processes in the nozzle with unsteady energy supply demonstrating the displacement of the nozzle shock wave towards the nozzle outlet in the over-expanded nozzle flow in comparison to perfect gas flow.

Chapter 3 considers integrals of 3D motion of a viscous and ideal medium. A study of relationships between integrals and areas of their applicability has been carried out for motion of incompressible medium. As a result of applying the methods of partial differential equations and mathematical physics it is shown that all considered integrals could be unite with the chain as a tree. On the base of a tree located the first integral of 3D Navier -- Stokes equations obtained by the author. This integral plays the role of a root integral. All other integrals united by the chain under consideration are its special cases. A proof is given that each of the well-known classical integrals of Bernoulli, Euler-Bernoulli and Lagrange-Cauchy is a special case of the root integral. So special cases of the root integral are six new ones obtained by the author and three well-known integrals of Lagrange-Cauchy, Bernoulli and Euler-Bernoulli.

Chapter 4 considers the movement in a deep water under action of gravity based on 3D Navier - Stokes equations for an incompressible medium. The main unknowns are pressure and three components of the velocity vector, while each of these quantities is a function of spatial coordinates and time. The density and kinematic viscosity of the fluid are assumed to be constant. The authors neglect an influence of all bounding surfaces with the exception of the free surface. Along the free surface they set the pressure constancy condition. The integral of 3D Navier - Stokes equations and the generator of solutions obtained by the author are proposed as the initial relations. As a result of the application of the methods of partial differential equations and mathematical physics the exact solutions of the Navier -- Stokes equations characterizing the movement in a deep water and near the free surface are found out. These results make possible to carry out a study of the nonlinear effects inherent in gravitational waves on the water surface. In particular the authors obtained the equations for free surface profile depending on coordinates, time and other governing parameters.

In: Navier-Stokes Equations ...
Editor: Peter J. Johnson

ISBN: 978-1-53619-967-3
© 2021 Nova Science Publishers, Inc.

Chapter 1

KINETIC MONISM AND ALL-UNITY IN RUSSIAN COSMISM VERSUS NEWTONIAN DUALISM OF SEPARATED ENERGIES

Igor Bulyzhenkov[*]

Moscow Institute of Physics and Technology, Moscow, Russia

ABSTRACT

The all-unity and kinetic monism of continuous matter in the Russian Cosmism contradicts conceptually the Newtonian dualism of kinetic and gravitational energies that ruined the intellectual legacy of many Russian thinkers. Lomonosov super-penetrating liquid, Umov inertial ether, Gurvish nonlocal biofields, Vernadsky Noosphere of thoughts, Chizhevsky cosmic pulsations, Bekhterev immortality of thoughts, Tsiolkovsky cosmic life and other orthodox suggestions are still incomprehensible and not in demand by the science textbooks and education curriculum based on negative (nonexistent) gravitational energies and the questionable "action-at-a-distance." Starting from the local cause of all accelerations according to Lomonosov and the hidden variable for extracting mass-energy from the inertial ether according to Umov, everyone can logically replace the fiction of gravitational potentials with non-local self-assembling of correlated densities in the volume integral of kinetic energy. Mathematical

[*] Corresponding Author's Email: bulyzhenkov.ie@mipt.ru.

superposition of continuous material densities, inaccessible to human perception and measurements, can deduce the inverse square law for mutual accelerations and revive many "weird" ideas of Russian cosmists. The kinetic cooperation of correlated densities and stresses of continuous mass-energy with absolute self-knowledge of its non-local distribution is a more fundamental component of cosmic human beings than their visual structure with the densest volumes in spatial communications. Self-control of adaptive kinetic densities sets in motion the material all-unity of nonlocal bodies and invisible distributions of thoughts over the Noosphere. The kinetic monism of nonlocal energy can verifiably modify Euler / Navier-Stokes fluids by inertial feedback and remove non-Euclidean 3-interval, empty space, gravitational energy, black holes, and dark matter from educational programs in physics, astronomy, biology, medicine, and other natural sciences.

Keywords: nonlocal reality, extended bodies, material space, metric self-assembling, adaptive Euler/Navier-Stokes flows, kinetic attraction and repulsion

INTRODUCTION TO MODERN CHALLENGES

In trying to separate genetics from unknown information-reading mechanisms, my brother called me one day to ask how physicists might explain that his dog repeatedly attacks TV images of lions and only lions. How could this Ridgeback Alma, born at the Geneva club, know that his ancestors were selected to fight African lions? At that time, I lectured at my class of General Relativity that metric spaces of Einstein can describe Cartesian matter-extension (Garber, 2000) even better than the empty space with Schwarzschild's singularities. In 2006, however, I was not ready to comment on the nonlocal mechanism for Noosphere thoughts recording/readings in the world material space. And, I replied with the on-duty quantum nonlocality of the invisible microworld, in the very terms one can find in "The Field" (McTaggart, 2002).

Professors of physics used to replace all unexplained mysteries of macroscopic phenomena by sophisticated fluctuations of quantum particles. Conventional interpretations from the Standard Model are our self-defense against unexplained (and rarely published in peer-refereed journals) challenges of practice or alternative philosophical conclusions.

Nowadays, Kant might not get a chance to share with the Web of Science journals that the mind has to possess the synthetic a priori or innate knowledge of "things in themselves" (Kant, 1781) because modern referees need to see this statement within the established causal science, rather than outside it. Because of conventional scientific dogmas, many original ideas of philosophers were unused by researchers in the 20-21 centuries.

Unexplained influences between distant events were already discussed by more than 65 world nations (McTaggart, 2002). Here my personal experience is related mainly to informal talks with extraordinary people, including scholars at the Russian Seminar for Time Nature Study (www.chronos.msu.ru). Unlike the natural sciences, philosophy was never considered by my university teachers in the mid seventeens to be a leading tool for the exploration of Nature. Professional physicists tried to ignore philosophical definitions like "matter is an objective reality given to us in perception" (the latter varies over humans and other live species). Nonetheless, motivated by the tremendous triumph of quantum physics for probabilities of delta-function events and 'observed' local collisions of elementary particles, modern scholars failed to suggest nonlocal physical options for human beings.

Contrary to top scholars, the multi-national population of the "materialistic" USSR in my youth (and in today's Russia) always believed in intuition, distant communications of native souls, and a cosmic nature of marriages. Gurwisch's biofields (Gurwitsch, 1944), Vernadsky's Noosphere (Vernadsky, 1945), Chizhevsky's pulsations of the Universe (Chizhevsky, 1976), Kozyrev's Causal Mechanics, Bekhterev's immortality of thoughts, Messing's psychic performances, Globa's astrology calendars, Vasiliev's parapsychology were pseudo-scientific only for the domestic Academy of Sciences but not for the monism and all-unity concept of Russian Cosmism (Young, 2012) and its followers (Tsiolkovsky, 1923, 1933; Roerich, 1990). As a native Russian, I can knowingly talk about the anti-Newtonian worldview only in the society of Russian cosmists. But, I am confident that many other nations have similar monistic ideas regarding Plato's material ether or Descartes' matter-extension which are still beyond the Standard Model of contemporary physicists. By revisiting the energy world monism of Russian Cosmism in relevant comments of Lomonosov, Umov, Tsiolkovsly and other domestic thinkers, I highly appreciated the concept of continuous material space-plenum of the Ancient Greeks. At the beginning of my study, I was not ready to look at the kinetic nature of Orthodox reality because my pro-Newtonian education did not allowed me to admit the pseudoscientific essence of negative gravitational energies in the physical reality of only positive kinetic energies and measurable values.

Following by the 'divine' distant pulls of Newtonian gravity (but not local pushes of Lomonosov's liquid-matter), the Russian Academy of Sciences and the Ministry of Science and Higher Education cannot satisfactorily comment the 'action-at-a-distance' nature between localized masses. Our educational textbooks widely employed the controversial negative (gravitational) energy for the 'atheistic reality' of only localized inertial bodies with retarded communications. In my today's view, reality should consist only on measurable (positive, kinetic) densities and their volume energies. The Yin-Yang dialectic of Chinese philosophy and the Orthodox mentality of Russian religious thinkers will never agree with an imposed combination of opposing complementary energies in one corporeal world of observations. Below, I will try to justify the urgent need to transfer textbooks on Classical Mechanics and Electrodynamics from the empty space of Newton to continuous densities of the material (kinetic) space of the Ancient Greeks and Descartes. New textbooks will be supported by the ideas of the cosmic life of classical German idealism (Hegel, 1807) and Russian Cosmism (Semenova, 1993; Hagemeister, M. 1997; Djordjevic, 1999). To reply the pro-Newtonian critics of the Russian monistic approach to the Universe all-unity (means nonlocality in micro, macro, and mega scales), I intend to call Newton's gravitation a pseudoscientific theory, since it relies on the fiction of non-existent (negative) energies-potentials. Recall that nobody has measured yet negative energy in the physical world of only positive energies. Therefore, the ongoing debate on how to quantize negative gravitational energies (rather than positive kinetic fields of inertia) seems completely unscientific from the point of view of misunderstood Russian cosmists.

MONISTIC METHOD

Since 1742, the first president of the Russian Academy of Sciences Mikhail Lomonosov tried to withstand the Newtonian concept of pulling gravitation in favor of local stresses of invisible matter-liquid due to the claimed monistic reason to charge any mechanical motion by kinetic pushes only (Lomonosov, 1950, 1970). Many European mathematicians were also against the Newton scheme of 'divine' distant interactions. Nicolas Fatio de Duillier first assumed the kinetic nature of pushing gravity in the 1690 letter to Huygens. Minute corpuscles of Fatio (Zehe, 1980) and ultra-mundane particles of Le Sage (Van Lunteren, 2002), both pushing locally upon visible bodies, did not convinced that time physicists, including Huygens. However, Bernoulli expressed an interest. And, Leibniz criticized Fatio's corpuscles only for demanding empty space between, which was denied by Leibniz (after Descartes) on philosophical grounds.

Contrary to the local pressure of Lomonosov's ideal liquid, mechanical corpuscles would warm probe bodies and damp rotation of planets in the steady Solar system. As a result, the dissipative ether of independent corpuscles did more harm than good for the kinetic theory of mutual attraction with the law of inverse squares.

In modern mathematics, the point mass of Newton moves in warped 3-space of General Relativity, while the point electric charge still moves in Euclidean 3-space. Where is logic in such a conformism? Do masses and electric charges exist in different Universes or is there a common sense in Newtonian physics of the empty space and point particles? Where could point-matter materialists fix cutting edges between observable inertial matter (an apple "given to us in perception") and its non-observable (beyond perception) parts, logically related to both observable and non-observable matter in the Universe of Kant-Hegel-Mach idealism? The substance surface cannot pass continuously through postulated empty space between postulated point material particles on the visual apple's edge. Once there is no matter in empty space beyond each atom or molecular, then there is no material line between two distant atoms (molecules), and there is no smooth material surface over the visible apple. Like there is no continuous material substances between the Earth and the Moon in the Newtonian textbooks of empty space proponents.

It is unlikely that Chinese dialectical teachers, Hindu philosophers, the Ancient Greeks, Descartes, German idealists, and (the most understandable for me) Russian cosmists had no logic. Thus, why and how did they all arrive at the 'weird' conclusions about invisible realities of the material world? Kant developed a deeper version of the celebrated Platonic forms. Continuous distributions of energies next to invisible substances over the Universe were employed in Chinese and Hindu philosophies well before Plato's dialogues regarding eternal forms "residing in material objects". Deductive logic, invented by Aristotle for a causal interpretation of Nature, had accepted that visual things have a lesser reality than Platonic forms. To Aristotle, substance is a combination of visible matter and some invisible forms, which are not a separate realm, nonetheless. He logically rejected the empty-space concept in favor of continuum space-plenum filled everywhere with a background of invisible things.

Starting from Plato's Dialogues, the Western classical idealism culminated in Hegel's dialectic method for Nature with the absolute self-knowledge and collective mind-spirit-soul, denoted by one German word *geist* (Hegel, 1807). Collective consciousness was coherently assumed by geochemist Vernadsky as the "sphere of human thought" (later called the Noosphere), which is the third phase of the Earth's formation after the Geosphere and the Biosphere (Vernadsky, 1945). Is it possible to understand what part of the Standard Model a modern astrophysicist may put behind Hegel's geist or Vernadsky's Noosphere? May one talk about de Broglie waves, zero point fluctuations, strings?

Physics teachers may surely talk about the invisible microworld and whatever they like, but philosophers discussed definitely monism and all-unity regarding the macro and mega worlds. Here idealists insist on the global superimposition of all minds in some information space with "the synthetic a priori knowledge." The mind is not separated from matter in such an approach to continuous space-plenum but can interact with distributed inertial matter through "transmutations of elements." In other words, the material mind and other matter (including the human body) should extend in the same interaction space with a spatial overlap between the collective mind and all human beings.

At first glance, this seems to be absolute nonsense, as everyone sincerely testifies that such a global superimposition of live or inert matter apparently contradicts our daily observations, say separated localization of different apples on a table. Does this mean that the Ancient Greeks, German idealists and Russian cosmists had neither logic nor practical experience in space science? Or that they, together with ballistic expert Mach (Mach, 1904) and physicist-cosmist of continuous energies Umov (Umov, 1874), did not understand classical mechanics for 'localized' apples? Furthermore, even Newton himself looked at the "absurd" ether idea for the material superimposition of his gravitating bodies. Faraday considered charged field-matter around a charge's 'center of force', Maxwell initiated the displacement current through continuous flows of charged matter, and Clifford speculated on inhomogeneous material space. Later Umov, Mie, Hilbert, Einstein, Born, Schwinger and others motivated never observed continuous (or extended) source of classical fields. Have all these outstanding professors lost logic for a moment? And, have proponents of only localized matter and retarded exchanges ever seen the amazing cohesiveness of bird flocks or fish schools when thousands of animals simultaneously correlate directions in timeless readings of what is going on?

MONISTIC MATTER-ENERGY OF RUSSIAN COSMISTS

Multi-Vertex All-unity of Continuous Energy

The monistic method to describe mechanical events through extended mass of shared material space (with local kinetic stresses instead of gravitational pulls between distant mases with localized kinetic energies) arises a reasonable question - who and when misinterpreted the kinetic reality, deductive logic, or observations of macroscopic volumes of continuous matter-space in daily practice? Quantum physics was irrelevant to the logical discovery of the Ancient Greeks that observations of (macroscopic) bodies have a lesser reality than their eternal forms that are beyond the level of human perception. To the Greeks, the real space-plenum, hidden from detailed observation, is filled by continuous matter with nonlocal bounds (now called correlations), which were re-examined in the Einstein-Podolsky-Rosen "spooky actions at a distance" (Einstein et al, 1935). Why not drop forever the unnecessary empty-space paradigm of Newton and the delta-operator source in Maxwell's and Einstein's field equations in favor of the ancient Greeks' continuous substance within the 'perception fog' of material space-plenum? Why not to follow the conclusion of ancient Hindu philosophers that a body (its visual frames, to be precise) is an illusion of our incomplete observations of extended cosmic matter? Why not to make apparent use from the Chinese dialectic of yin-yang worlds for negative and positive energies? All these findings can be rigorously employed by Maxwell's Electrodynamics (ME) and Einstein's General Relativity (GR) with analytical particle densities (Bulyzhenkov, 2008, 2018) rather than with the Dirac delta-densities.

Early or later, academic researchers will agree that it is absolutely illogical to maintain that substance in the microworld is nonlocal and should be described in monistic terms of material fields, but localized substance in the macroworld is not a field distribution. No one else found a spatial scale in doubt for transition between nondual microscopic and dual macroscopic states of inertial mass-energies. Restricted frames of observed macroscopic bodies are perception illusions formed by concentrated bunches of dense vertices of r^{-4} radial elements of inertial matter extending over the entire Universe.

Einstein's equation of 1915 states that the elementary source of inverse square fields, $f \sim 1/r^2$, is the distributed energy-tensor density, $f \bullet f \sim 1/r^4$, rather than the point mass. From here on, each mechanical body (or carrier of mass-energy in modern relativistic equations) should be considered as the multi-vertex continuous distribution of inertial mass-energy in the monistic world continuum. Therefore, Einstein's sources of metric fields directly addresses extended masses with inhomogeneous metric stresses as in Lomonosov's liquid-matter, Aristotle's space-plenum with invisible continuous things, and energy-matter flows in Hindu and Chinese philosophies of the ancient East. Restricted frames of observed macroscopic bodies are perception illusions formed by concentrated bunches of dense multi-vertex fields of inertial matter (extending over the entire Universe). The inverse bi-quadratic mass-energy density $\sim 1/r^4$ of the elementary space-matter distribution arranges locally the $\sim 1/r^2$ metric accelerations to attract dense volumes of probe bodies in such weak fields, while ultra-strong mass-energy densities repeal probe bodies (Bulyzhenkov, 2018a).

It is clear from the inverse bi-quadratic distribution of an extended particle that when observers cut the vertexes of continuous bodies over 'the clouds of human perception' (to approximate practice by localized masses), the distant gravitation in empty space of Newton inevitably contradict to the monistic self-organization of nonlocal space-matter with the solely kinetic content for inertial mass-energy. The superficial model of localized bodies can hardly be agreed with thoughts of ancient and contemporary cosmists, Descartes, and Hegel, the recognized father of the universal dialectical method. The most of materialists (and many contemporary researchers) associate the ultimate reality to only detectable energy integrals that are above the perception level. Radial densities (invisible but still material) or fine continuations of superimposed radial particles do not fall under the materialistic definitions of matter.

If mechanical bodies were indeed localized formations in empty space of distant interactions rather than in formal mathematical models with unphysical infinite energies of material points, then the definition of substance through finite human perception might make some educational sense. But, non-empty space organization of continuous matter exists below the restricted human perception. Invisible material continuations with low periphery densities of volume integrals mc^{-2} are even more fundamental for the nonlocal world organization than its dense volume regions, called bodies. For example, invisible material thoughts in the Vernadsky's Noosphere and the Kant-Hegel's Universe with the absolute self-knowledge of Nature continue to maintain steady energy-information forms even after the full disintegration of macroscopic bodies of former thinkers into gases of atoms.

If there is a universal tool for writing Noospheric thoughts in a non-empty spatial continuum of inertial and electromagnetic energies, then a complimentary mechanism for reading this information from the Noosphere must also exist in such all-unity of superimposed elements with the global storage of data. The supposed mechanism for reading / writing material thoughts of the Noosphere may shed some light on why the dog Alma has repeatedly attacked TV images of lions that never before. Such a reading of subtle information levels in the common material space of the population all-unity of extended dogs and lions is associated with the recognition abilities of consciousness, not with gene codes.

The information reading mechanisms in the nonlocal world self-organization can justify not only intuition, but also the known practices of remote viewing (RV) and information-driven phenomena of non-traditional medicine. Homeopathy, parapsychology and Chinese medicine cannot be called pseudoscience based solely on conventional Newtonian textbooks with dual (field + substance) models of mechanical bodies localized in empty space. Again, Newton's mechanics is based itself on the fiction of nonexistent gravitational potentials and, therefore, is a vivid example of successful pseudoscience (similar to the very successful pseudoscience of epicenters of Claudio Ptolomeo in the geocentric paradigm).

Monism of Continuous Mass-Energy with Correlated Kinetic Stresses

The author is not against Ptolomeo or Newton pseudoscience models with many successful predictions, but against aggressive suppression of alternative descriptions of nature by cosmists and idealists within the framework of available measurements. Western scholars tend to relate Einstein's geometrical space-time with the Newtonian palliative of the point particle and unphysical delta-operators in energy integrals of material densities. Following the West, modern Russian physicists and mathematicians also ignore the advanced concepts of Russian Cosmism regarding energy monism and all-unity of continuous world densities. However, the empty space of Newton and the notion of negative gravitational energies are in intellectual contradiction with Aristotle's logic, Descartes' matter-extension, Umov's inertial flows of material ether (Umov, 1874), and the positive kinetic energies $mc^2\sqrt{g_{oo}}/\sqrt{(1-\beta^2)} > 0$ of Einstein in any metric fields. The negative field potential of Newton seems me suitable only to the incorporeal (yin) world in Chinese dialectic but not the corporeal (yang) physical reality of positive (kinetic) rest-energies $mc^2\sqrt{g_{oo}}$.

The nonlocal superimposition and self-organization of kinetic mass-energies of inertial material spaces of inanimate bodies and extended human beings seems contradictory only to observations with a finite perception level. This 'weird' overlap of all continuous substances, logically inferred by idealists, does exist in material space organization for Einstein's metric mechanics and Maxwell's electrodynamics. Here the nonlocal carrier of elementary mass-energy integral, $mc^2 \equiv r_o c^4/G$, can be analytically described (Bulyzhenkov, 2008) by the continuous radial density $mc^2 r_o /4\pi r^2(r+r_o)^2$ instead of the delta - density $mc^2\delta(r)$ for the controversial model of localized particles in curved 3-space.

Analytical mathematics of continuous masses and charges (Bulyzhenkov 2008, 2012, 2018) maintains the global spatial superimposition (of all extended sources and their classical fields in the world material space of Descartes) behind the Ancient Greeks' logic and Kant's interpretation of Platonic forms. Euclidean space 3-geometry (which is the Kant example of *a priori* knowledge in Nature) matches Machian nonlocality of continuous masses (Bulyzhenkov, 2008) and Kant cosmology of steady cosmic organizations. Mach indeed had analytical grounds for overlapping macro-states (Mach, 1904) of invisible inertial matter and for logical criticism of localized Boltzmann atoms. The true nature of 'microscopic' corpuscles in Einstein's physics corresponds to the r^{-4} radial sources distributed over the infinite Universe. Therefore, the relativistic mechanics of continuous mass-energies should depend on their spatial overlap and corresponding correlations with all 'distant' radial stars or with the Machian "rest of the Universe."

The r^{-4} distribution of one elementary mass and their nonlocal multi-vertex system can be rigorously described (Bulyzhenkov, 2018) through the self-consistent continuum of monistic energy in Maxwell's electrodynamics and Einstein's metric formalism, rather than through allegedly localized bodies. Researchers and physics teachers fairly ought to admit that ancient philosophers and contemporary cosmists interpret real continuous bodies and the r^{-4} nonlocality of each human much better than the leading laboratories with the mega-science infrastructural facilities. Newtonian proponents still associate macroscopic sources of electric and gravitational fields with localized bodies or systems of point particles. The universities in Russia (and other countries) has not yet appreciated the Mie (Mie, 1912) and Einstein-Infeld (Einstein, 1938) directives toward extended sources in the nondual theory of classical fields. But, the classical field equations self-consistently accept the elementary radial particle integrated into the very spatial structure of its fields (Bulyzhenkov 2008, 2009). The corresponding global superimposition of all inanimate and living bodies was logically inferred by many Russian cosmists, including those who declared the coexistence of mind and body in the same material cosmos (Semenova, 1993; Hagemeister, M. 1997; Djordjevic, 1999).

The extended energy-charge can by justified by Einstein's metric formalism with the Christoffel connections for local geometrical stresses in the multi-vertex continuum of kinetic densities (Bulyzhenkov, 2018). The nonlocal sub-organization of elementary mass-energy $m_k c^2$ = const in a multi-vertex world system forms kinetic distribution on a lower hierarchy level. The volume integral of non-locally correlated stresses around each elementary vertex defines its geodesic acceleration, which Lomonosov associated logically with local pushes of invisible liquid-matter (Lomonosov, 1742). Metric stresses of continuous space-matter can accelerate the probe mass for local measurements in the laboratory. 'Divine action at a distance' is not a consequence of nonexistent gravitational potentials. The phenomenon of gravity originates from non-equilibrium local pushes of adaptive kinetic densities under their nonlocal correlation in a closed system of continuous mass-energy.

The author is not in a suitable position to describe what was happening in foreign countries with cosmic and monistic ideas of their philosophies. In Russia, there are facts of two 'philosophical steamboats' for departed 'idealists' in 1922, followed by expels of other thinkers to the Main Administration of Corrective Labor Camps and Settlements (GULAG). In the absence of a national committee for verification of unexplained experimental data, the Russian Academy of Sciences is still fighting with the anti-Newtonian worldview of Russian Cosmism and the Lomonosov theory of local ether stresses. Instead of the invisible ether-matter of Lomonosov and Umov with a hidden degree of freedom for controlled extraction of kinetic energy, the national scientific administration is trying to rely on non-physical black holes, non-Euclidean and empty space, dark matter models and similar adventures of Newton-based theorists against the legacy of exiled Russian philosophers. And, the Russian Commission to Combat Pseudoscience and Falsification of Scientific Research considers the negative gravitational energy, curved 3-space, the black hole and dark matter existence to be firmly established scientific facts. However, Einstein knowingly denied metric singularities for the physical reality since 1939. Russian cosmists never criticized Einstein's geometrization of physics due to the monistic options to unify extended matter with metric stresses of space-time.

DISCUSSION

Metric Stresses in General Relativity for Local Pushes of Lomonosov instead of Distant Gravitation Pulls of Newton

The author did not meet a single one academician who publicly agrees with Kuhn - "Einstein's theory can be accepted only with the recognition that Newton's was wrong" (Kuhn, 1962). Everyone is inclined to believe that Einstein's mechanics is a bit more complex version of Newton's with the same ontology of inertial masses and gravitational fields. However, Newton's description of warm bodies ignores Einstein's rest energy mc^2

or the variable content kmc^2 of inward energy, which Russian cosmists associated with ether parts of massive bodies starting from 1873 (Umov, 1874). It became clear that Newtonian dynamics of point masses contradicts the tensor self-organization of GR densities under competing thermal chaos and translation order in moving distributions. The collinear relation between applied vector forces and observed accelerations (the second law of Newton) works only for volume integrals of densities, but not for densities themselves. The Umov variable for the unmeasurable (inward) mass-energy density $k\mu c^2 \leq \mu c^2$ was identified with the Lagrangian (up to the opposite sign) next to the unmeasurable kinetic term (Hamilton's double energy pv) in the Lorentz transformations. The true, monistic nature of nonlocal distributions of only positive, thermo-kinetic heat-energy is misinterpreted by textbooks on Newtonian mechanics for measurable energies ($pv/2 = \mu v^2 /2$) through the invented negative potential (which is a nonexistent energy or a fiction).

Contrary to Newton's gravitation, Einstein's metric theory operates only with positive (kinetic) mass-energy that matches the monistic nature of nonlocal energy organizations. The phenomenon of distant interactions between dense (visible) regions is not a reason for their motion but a direct consequence of the nonlocal energy distribution with correlated metric stresses. Pseudo-scientific negative energies are inherent only to dual approaches to the 'divine' gravitation of Newton. Physics teachers should introduce this dual theory only as a palliative simplification of non-local monistic reality with positive (kinetic, measurable) energy densities and positive volume integrals.

One more time, there are no negative gravitational energies at all in the monistic reality of continuous cosmic all-unity in Russian Cosmism. The error of Newton, just to answer Kuhn on behalf of Russian cosmists and yin-yang thinkers, if admissible to the author, was that positive (yang) kinetic energies were mixed with negative (yin) gravitational energies in the dualistic mechanics of localized substances and delocalized forces-fields. The negative gravitational potential, assigned mathematically to the observed phenomenon of gravity, is not the true reason for local accelerations. This potential is not a reason but a calculation consequence of nonlocally correlated kinetic densities due to their integral conservation for different spatial distributions.

Energy monism for the all-unity of continuous cosmic densities in Russian Cosmism philosophy and Yin-Yang complementary of worlds in many Eastern philosophies forbid mixing of corporeal (kinetic) and incorporeal (non-kinetic) notions for the physical reality of positive energies. Plato in the Timaeus says that "matter and space are the same". Spatial energy distributions and matter are also the same in quantum theories since 1923. In 1938, Einstein and Infeld proposed to describe monistically matter and energy fields in the same metric terms of General Relativity. Why does the Ministry of Science and Higher Education persistently teach in national universities and high schools the dual pseudoscience of Newton instead of the monistic all-unity of Lomonsov, Umov, Tsiolkovsky and other Russian cosmists? Why aren't Chinese universities developing yin-yang physics and relevant mathematics for the well-known phenomena of non-local being and alternative medicine?

No Dark Matter in Monism and All-Unity of Kinetic Densities

The search for dark matter will not make much sense if researchers accept the energy monism and all-unity (nonlocality) for all microscopic, macroscopic and mega-scale distributions. Einstein's metric space-matter with extended mass (Bulyzhenkov, 2008) and with local stresses-pushes of Lomonosov (Lomonosov, 1742) in continuously distributed kinetic densities presupposes their correlation and self-control over the constant integral of positive energy of an almost isolated system (like a free elementary particle or galaxy).

Static self-organization of elementary mass-energy around one vertex leads to a radial self-distribution of inertial densities with an equilibrium metric solution in a flat material space without singularities or black holes (Bulyzhenkov, 2008). Multi-vertex solutions must exhibit continuous crystal structures with metric wave-matter or sound capabilities. The exchange of wave energy between dense regions of an isolated distribution and its periphery can support a macroscopic tornado or a spiral galaxy in accordance with the universal scenario for all material continua. No one looks for dark matter profiles in the Ranque-Hilsch vortex tube. Likewise, a spiral galaxy must be revisited due to the nonlocal self-organization of kinetic energies with internal heat transfer or inelastic wave exchange. Monistic self-organization of continuous tornadoes with losses of kinetic energy of the central regions due to outgoing waves should mathematically describe the rotation curves of a spiral galaxy without the need for dark matter. Conceptually, Russian Cosmism encourages readers to develop theories of self-rotation and self-pulsation of continuous cosmic densities instead of dark matter profiles.

Umov's Energy Media with Tensor Self-Organization of Correlated Densities versus Euler/Navier-Stokes Transfer of Point Masses

The Navier-Stokes corrections for the Euler hydrodynamics of passive point masses correspond to the straight-line counteraction to the applied vector forces. However, such anti-collinear inertial responses are applicable only for test bodies whose masses are volume integrals of their densities. In contrast to the vector currents of mass integrals, continuous densities have a tensor organization. The inertial response of tensor densities can be directed both parallel and perpendicular to the local vector gradients of the external pressure. Such a multidirectional reaction of Umov's continuous energy to accelerating force densities (not their integrals) can explain the enhanced transfer of thermal energy perpendicular to jet streams and plasma currents in Tokamak reactors.

Many industrial systems with non-uniform energy flows cannot be satisfactorily described by the Newtonian vector action-vector reaction mechanism. The all-unity (nonlocal) approach of Russian cosmists to monistic kinetic densities makes it possible to modify the Euler / Navier-Stokes equation using the "vector action - tensor reaction" for adaptive energy distributions. This formalism or all-unity for continuous material space allows self-organization of its volume regions, their auto-pulsations, and clearly predicts turbulent regimes of moving densities, in contrast to Newtonian straight-line accelerations of energy integrals.

In any case, the one-line Newton-Euler reaction does not work for a three-dimensional continuum of GR tensor distributions. Consequently, the vector transfer of passive masses in the Euler / Navier-Stokes equation should be modified by metric stresses in all three directions, if we accept the all-unity of kinetic densities and metric fields instead of the dualism of spatially separated gravitational fields and kinetic singularities. Again, the true cause for the spatial motion of visible bodies is not negative gravitational attraction from distant bodies, but local metric stresses in the correlated distribution of kinetic densities. The equipartition principle for kinetic energies at external and internal (ether) degrees of freedom controls the local accelerations / decelerations of Lomonosov in the metric self-organization of extended matter (Bulyzhenkov, 2018).

Kinetic Monism of Self-Pulsating Cosmic Organizations and the Accelerated Metagalaxy with Self-Cooling

There is no need at all for negative (gravitational) energies in the monistic approach to continuous pushing energies of the Russian cosmists (Lomonosov, Umov, Tsiolkovsky, Vernadsky, Chizhevsky, Gurwitsch, Bekhterev, Kaznacheev, et al.). The all-unity of internal (ether, unmeasurable rest-energy) and external (extracted, measurable) contributions to the system integral of kinetic energies, $\Sigma m_k c^2 = $ const, controls both metric attractions and metric repulsions of dense inertial volumes. This monistic self-organization of continuously distributed kinetic energies contradicts to the assumed tendency to reach the lowest gravitational potential in false gravi-mechanics of localized masses.

There is no Newton's law of universal gravitation in the teachings of Russian Cosmism. There is a law of all-unity of monistic matter, which assumes dynamic pulsations and other non-local self-organization of continuous densities. The complementary motion of chaotic (ether) and ordered kinetic energies allows using the principle of equipartition (Bulyzhenkov, 2018) for metric fall and take-off in the relativistic physics of dense material space. General Relativity, based on the metric references for continuous radial mass without metric singularities, can explain all known gravitational experiments under Euclidean 3-geometry in pseudo-Riemannian space-time (Bulyzhenkov, 2012). The structural self-organization of this material space with mutual oscillations of Umov's ether chaos and the kinetic order of measurable translations is not described by Newtonian mechanics, where masses do not have the hidden ether variable $kmc^2 \leq mc^2$ of Russian cosmists.

The Megagalaxy can pulsate around the equipartition of internal energy-chaos and ordered kinetic energy, like any other oscillating self-organization in the world kinetic hierarchy. The modern stage of the Hubble expansion with coordinate acceleration means a rapid decrease in the relativistic physical velocity of scattering parts and the concomitant cooling of their material space in the monistic physics of Russian cosmists. The quantitative relations for the oscillating self-motion of elastic organizations can be analyzed in metric terms of General Relativity (Bulyzhenkov, 2008, 2018a).

Self-oscillations of density shapes in non-local energy distributions, including periodic auto-pulsations of a non-local Metagalaxy, occur at a constant mass integral of the whole system $\Sigma m_k =$ const or a constant kinetic energy of this system. There is no need for dark energy to describe the scattering phase with accelerations, since the multi-vertex mechanical distribution takes into account the well-defined interference contributions of two-vertex densities to the continuous mass-energy of space-matter $Mc^2 = \Sigma m_k c^2$.

CONCLUSION

Einstein's metric theory of space-time can fix Newton's error by incorporating negative (gravitational) shifts into always-positive values of kinetic energies. This kinetic approach with the rest mass-energy stroked out negative gravitational energies from the metric theory of motion. There are both attracting and repulsing self-tensions of distributed kinetic energies under the nonlocal self-organization of multi-polar inertial systems (Bulyzhenkov, 2018). Assuming the kinetic nature of the monistic physical reality, metric interactions in the non-equilibrium material space is better called the theory of continuous inertia, and not (nonexistent) gravity. If the "philosophical ships" did not send top cosmists from Russia in 1922, and then the authorities did not repress Russian religious idealists, then the pushing liquid-matter of Lomonosov and the Umov dynamical variable $k(v)mc^2 \approx [1 - (v^2/2c^2)]mc^2$ of the inward (ether) heat-energy could together clarify the kinetic nature of Einstein's metric accelerations in domestic textbooks many years ago.

The shared material space of extended masses with inertial flows of thermal energies coherently rearranges Einstein's physics through the continuous elements of metric inertia in the monistic line of cosmic being. Without Newtonian pseudo-gravity, there would be quantitative laws for describing a nonlocal biofield according to Gurwitsch, the material Noosphere according to Vernadsky, cyclic pulsations of material space according to Chizhevsky, cosmic life formations according to Tsiolkovsky, the immortality of Noosphere thoughts according to Bekhterev, etc. Aristotle logic and philosophical conclusions should dominate in the metaphysics of Nature over materialistic pragmatism, incomplete observations and operator-based mathematical approximations of reality.

Nonlocal correlations of bio-cells (Kaznacheyev, 1976), bound photons (Aspect et al, 1982), distributed electric charges (Emelyanov, 2018), polarized quasars in large-scale structures (Hutsemekers et al, 2014) and even coherent galaxies (Lee et al, 2019) have already been proven by recognized scientific methods. The extended mind and collective consciousness are successfully explored in many research programs on the Noosphere. Many data have already confirmed the cosmic nature of living matter and the nonlocal correlations of distributed mass-energies.

In such a monistic world of self-governed kinetic energies, everyone is truly a citizen of the entire Universe, not only Earth (Tsiolkovsky, 1923). In short, the educational curriculum should accept the material space of continuous inertia in macroscopic physics and replace the elementary current $e_k\, \delta(x - r_k)\, dr_k/dt$ in the Maxwell-Lorentz equations with the continuous density j(x) for the distributed charge integral e_k at any material point x. The author urges researchers to reconsider the point source in favor of the non-empty space of continuous mass-energy with local vertexes. Metric space-matter of closed mechanical systems is responsible monistically for the local metric stresses and for nonlocal correlations of metric energy densities around vertexes. Everything, including thoughts, are continuous formations of kinetic sub-energies in the continuous cosmic all-unity of organic and inorganic densities.

Nonlocal kinetic energy of interaction space-matter is to be a primary notion in all science disciplines. The monistic primacy of kinetic energy might reconcile one day materialists and idealists. Cosmic consciousness in Hinduism and Buddhism, Chinese dialectics, German classical idealism and the ultimate reality of material thoughts and cosmic human beings in Russian Cosmism should be mathematically revisited with the kinetic formalism for Einstein's metric fields in terms of Cartesian matter-extension. The kinetic hierarchy of multi-vertex energy continua and the correlated all-unity of all material densities with nonequilibrium local stresses can finally reject a nonexistent gravitational energy and 'divine' distant pulls from the educational programs on relativistic mechanics and cosmology with 1873 Umov's variable $\sqrt{1-\beta^2}\mu c^2$ for the extractable energy density of adaptive ether.

References

Aspect, A, Dalibard, J. L., and Roger, G. (1982). "Experimental Test of Bell's Inequalities Using Time - Varying Analyzers," *Phys. Rev. Let.* 49, 1804-08.

Bulyzhenkov, I. E. (2008) "Einstein's Gravitation for Machian Relativism of Nonlocal Energy-Charges," *Int. J. Theor. Phys.* 47, 1261-1269. https://doi.org/10.1007/s10773-007-9559-z.

Bulyzhenkov, I. E. (2009). "Superfluid Mass-Energy Densities of Nonlocal Particle and Gravitational Field." *Jour. Supercond. Nov. Magn.* 22, 723-727, https://doi.org/10.1007/s10948-009-0583-5.

Bulyzhenkov, I. E. (2012). "Geometrization of Radial Particles in Non-Empty Space Complies with Tests of General Relativity". *Journal of Modern Physics,* 3 No. 10, pp. 1465-1478. doi: 10.4236/jmp.2012.310181.

Bulyzhenkov, I. E. (2018). "Cartesian Material Space with Active-Passive Densities of Complex Charges and Yin-Yang Compensation of Energy Integrals," *Galaxies* 6(2), 60, https://doi.org/10.3390/galaxies6020060.

Bulyzhenkov, I. E. (2018a). "Gravitational attraction until relativistic equipartition of internal and translational kinetic energies." *Astrophysics and Space Science* 39, 363. https://doi.org/10.1007/s10509-018-3257-6.

Chizhevsky, A. L. (1976). *"Earth echo of solar storms."* (Moscow, Thought, in Russian).

Djordjevic, R. (1999). "Russian Cosmism," *Serb. Astron. Jour.* 159, 105-109.

Einstein, A., Infeld, L. (1938). *"The Evolution of Physics,"* (Cambridge Press, Cambridge).

Einstein, A., Podolsky, A., and Rosen, N. (1935). "Can Quantum-Mechanical Description of Physical Reality Be Considered Complete?", *Phys. Rev.* 47, 777.

Emelyanov, S. A. (2018). "From Relativistic to Quantum Universe: Observation of a Spatially-Discontinuous Particle Dynamics beyond Relativity." *Universe* 4, 75. https://doi:10.3390/universe4070075.

Garber, D. (1992). *Descartes' Metaphysical Physics* (University of Chicago Press, Chicago).

Gurwitsch, A. G. (1944). *"The Theory of the Biological Field"* (Sovetskaya Nauka, Moscow).

Hagemeister, M. (1997). "Russian Cosmism in the 1920s and Today" pages 185-202 in Bernice G. Rosenthal (ed.). *The Occult in Russian and Soviet Culture* (Cornell UP, Ithaca, London). ISBN 0-8014-8331-X; http://en.wikipedia.org/wiki/Russian_cosmism,

Hegel, G. W. F. (1807). "Phenomenologie des Geistes"; *Enzyklopadie der philosophischen Wissenschaften,* 3rd ed. 1830, (Hegel's Philosophy of

Mind, tr. William Wallace, 1894; *Phenomenology of Mind*, tr. J. B. Baillie, 1910; *Hegel's Phenomenology of Spirit*, tr. A. V. Miller, 1977).

Hutsemekers, D., Braibant, L., Pelgrims, V., and Sluse, D. (2014). "Alignment of quasar polarizations with large-scale structures", *Astronomy and Astrophysics* A18, 572.

Kant, I. (1781). "Critique of Pure Reason," trans. Norman Kemp Smith (N.Y.: St. Martins, 1965), A 51/B 75; Oizerman, T. I. Kant's Doctrine of the "Things in Themselves" and Noumena, *Philosophy and Phenomenological Research* 41, 333-350 (1881).

Kaznacheyev, V. P. (1976). *Psychological Systems* 1, 141.

Kuhn, T. S. (1962). "*The Structure of Scientific Revolutions*" (Chicago).

Lee, J. H., Pak, M., Song, H., Lee, H.-R., Kim, S., and Jeong, H. (2019). "Mysterious Coherence in Several-megaparsec Scales between Galaxy Rotation and Neighbor Motion," *Astrophysical journal* 884(2), 104.

Lomonosov, M. V. (1950). Complete Works, 11 Vols., *"Notes on the severity of bodies,"* 1743-1744, Vol. 2 (eds. S. Vavilov and T. Kravetz (Akad. Nauk. SSSR, Moscow and Leningrad).

Lomonosov, M. (1970). Editor Henry M. Leicester, "On the Relation of the Amount of Material and Weight", *Mikhail Vasil'evich Lomonosov on the Corpuscular Theory* (Cambridge: Harvard University Press), pp. 224 - 233.

Mach, E. (1904). "Die Mechanik in ihrer Entwickelung historisch-kritisch dargestellt," ["*Mechanics in their development presented historically and critically.*"] S. 236. F.A. Brockhaus, Leipzig.

Mie, G. (1912). "Grundlagen einer Theorie der Materie," ["Basics of a theory of matter."] *Ann. der Physik* 37, 511-534 (1912); 39, 1-40 (1912); 40, 1-65 (1933).

McTaggart, L. (2002). "*The Field: the Quest for the Secret Force of the Universe,*" (HarperCollins Publishers, New York).

Roerich, S. (1990). Testamentary disposition "*Archives and Heritage of Roerich for the Soviet Roerichs Foundation.*"

Salart, D., Baas, A., Branciard, C., N. Gisin, N., and Zbinden, H. (2008). "Testing the speed of spooky action at a distance," *Nature* 454, 861-854.

Semenova, S. G., Gacheva, A. G. (1993). *"Russky Kosmism,"* (Pedagogika Press, Moskva).

Tsiolkovsky, K. (1923). *"The Cosmic Philosophy"* (in Russian); *"Citizens of the Universe",* 1933 (in Russian). http://tsiolkovsky.org/en/the-cosmic-philosophy/

Umoff (Umov), N. A. (1874). "Beweg Gleich. d. Energie in contin. Korpern, Zeitschriff d. Math. and Phys." ["Move right away. d. Energy in contin. Corpses, magazine d. Math. And Phys."] V. XIX, Schlomilch. Umov, N. A., *"Selected works"* (Moscow - Leningrad, 1950, in Russian).

Van Lunteren, F. (2002). "Nicolas Fatio de Duillier on the mechanical cause of Gravitation," in Edwards, M.R. (ed.), *Pushing Gravity: New Perspectives on Le Sage's Theory of Gravitation* (Montreal: C. Roy Keys Inc.) pp. 41–59.

Vernadsky, V. I. (1945). "Some Words about the Noosphere", *The American Scientist,* Jan., 1945. https://21sci-tech.com/translations/The_Noosphere.pdf.

Young, G. M. (2012). "The Russian Cosmists: The Esoteric Futurism of Nikolai Fedorov and His Followers," *Oxford Scholarship Online.* doi:10.1093/acprof:oso/9780199892945.001.0001.

Zehe, H. (1980). "Die Gravitationstheorie des Nicolas Fatio de Duillier." ["Nicolas Fatio de Duillier's theory of gravity."] *Archive for History of Exact Sciences*, Hildesheim: Gerstenberg, 28 (1), pp. 1–23.

In: Navier-Stokes Equations ...
Editor: Peter J. Johnson

ISBN: 978-1-53619-967-3
© 2021 Nova Science Publishers, Inc.

Chapter 2

SIMULATION OF HIGH-TEMPERATURE FLOWS IN NOZZLES WITH UNSTEADY LOCAL ENERGY SUPPLY

N. Brykov[1], V. Emelyanov[1] and K. Volkov[2,]*

[1]Baltic State Technical University, St Petersburg, Russia
[2]Faculty of Science, Engineering and Computing, Kingston University, London, United Kingdom

Abstract

A model for the numerical study of unsteady gas dynamic effects accompanying local heat release in the subsonic part of a nozzle for a given distribution of power of energy release is developed. The finite volume method is applied to solve unsteady compressible Navier-Stokes equations with high-temperature gas effects. Solutions of some benchmark test cases are reported, and comparison between computational results of chemically equilibrium and perfect air flowfields is performed. The constructed model is applied to flows with unsteady energy supply in nozzles, which are of interest for alternating current plasmatrons. Moving arcs are formed which are sources of intense energy supply. It is allowed to burn one or more arcs that are at different points in space, have different intensities and move at a given speed relative to the flow. A qualitative structure of shock wave and thermal processes in the nozzle

---

[*]Corresponding Author's Email: k.volkov@kingston.ac.uk.

during unsteady energy supply is discussed. The results of numerical simulation of one-dimensional and two-dimensional under- and overexpanded nozzle flows with a moving region of energy supply are presented. Output nozzle parameters are calculated as functions of a number and time of burning of plasmatron arcs. The results obtained show a qualitative pattern of gas dynamics and thermal processes in the nozzle with unsteady energy supply demonstrating the displacement of the nozzle shock wave towards the nozzle outlet in the over-expanded nozzle flow in comparison to perfect gas flow.

Keywords: computational fluid dynamics, real gas, nozzle, plasmatron, energy supply, shock wave

1. INTRODUCTION

Gas flows in channels with a variable cross-sectional area with unsteady local energy supply arise in alternating current plasmatrons, high-voltage lightning protection switches and other technical devices. In connection with requirements of practice aimed at reducing the number of tests of designed products and the terms of development, there is an increased interest in the mathematical modelling and simulation of flows in nozzles with intensive energy supply [1].

The effect of spatial and temporal characteristics of a pulsating energy source on the supersonic flow in an expanding channel is considered in [2]. The use of low-frequency energy sources to improve propulsion characteristics produces higher values of specific impulse than a continuous supply of energy.

The transition from a supersonic flow regime in a channel to a subsonic one is accompanied by an increase in static pressure. The interest in the pseudo-combustion mode is due to the fact that during its organization, intensification of the mixing of fuel with an oxidizer occurs, and the rates of chemical reactions increase (due to an increase in pressure and temperature). Controlling the position of the pseudo-shock wave presents a considerable difficulty both in the isothermal flow and in the organization of combustion.

Conventional control methods based on stabilizers (ledges and pylons) lead to total pressure losses, and the position of the pseudo-shock wave is attached to the elements of the stabilizers. The possibilities and methods of controlling a pseudo-shock wave in a smooth channel of constant cross section with a pulse and periodic energy supply are shown in [3]. Design of Laval nozzle for real gas flows is discussed in [4].

The results of calculations of unsteady quasi one-dimensional flow in a channel representing an element of a ramjet engine are reported in [5]. The influence of the parameters of the energy supplied in the pulse and periodic mode (power, pulse frequency, distribution of sources along the channel length) on the flow characteristics is determined.

The behaviour of the electric arc in a supersonic nozzle flow is studied in [6, 7]. The energy supply leads to the displacement of the nozzle shock towards the inlet section of the channel compared to the flow without an energy supply and to the formation of vortex structures deforming the shape of the nozzle shock. The Prandtl turbulence model and the $k-\varepsilon$ turbulence model are used to close the Reynolds averaged Navier–Stokes (RANS) equations.

A two-dimensional model of flow in a plane and axisymmetric nozzle with local gas heating in the supersonic part, based on the Navier–Stokes equations and equations describing the heat balance in a gas heated by laser radiation, is developed in [8]. Parameters of the gas flow and the volume energy input, at which the gas heating does not exceed 1000–2000 K, are studied. The localized energy input into the flow is accompanied by an increase in pressure in front of the energy release region. When the energy input is sufficiently large, a shock wave occurs, which is located upstream of the energy release region.

A method of numerical simulation of the internal flows of a viscous gas, taking into account non-equilibrium chemical processes and the equilibrium excitation of the internal degrees of freedom of molecules, is developed in [9]. A flow of chemically non-equilibrium gas in a plane channel with a variable cross-sectional area is considered (a five-component model of air is used). The Euler and Navier–Stokes equations for the simulation of the development of zones of local energy supply in supersonic air flow in a channel are used in [10]. In the model of a real gas, the changes in the shock wave structure and flow parameters in the vicinity of the energy supply zone developing in a stationary medium and in the conditions of its interaction with a normal shock at different energy supply intensities, are discussed. High-temperature flows in nozzles are discussed in [11] based on the results of numerical simulation.

In this study, a model for the numerical study of unsteady gas dynamic effects accompanying local heat release in the subsonic part of a nozzle for a given distribution of power of energy release is developed. The constructed model is applied to flows with unsteady energy supply in nozzles, which are of interest for alternating current plasmatrons. Moving arcs are formed which are sources of intense energy supply. It is allowed to burn one or more arcs that are at different points in space, have different intensities and move at a given speed relative to the flow. A qualitative structure of shock wave and thermal processes in the nozzle during unsteady energy supply is discussed. The dependence of the nozzle output characteristics, as well as the displacement of the nozzle shock wave in over-expanded nozzle flow, on the intensity and cyclical energy supply in the subsonic part of the nozzle is reported.

2. MATHEMATICAL MODEL

In simulation of flows with energy supply, the problem is divided into a gas dynamic problem with given heat sources and a physical problem, in which the mechanism of heat release is studied for known flow parameters. The gas dynamic problem is solved using the Euler equations (the viscosity and thermal conductivity of the medium are ignored) or using the Navier–Stokes equations (for a viscous heat conducting gas).

2.1. Navier–Stokes Equations

In Cartesian coordinates (x, y, z), the unsteady viscous compressible flow is described by the equation

$$\frac{\partial Q}{\partial t} + \frac{\partial F_x}{\partial x} + \frac{\partial F_y}{\partial y} + \frac{\partial F_z}{\partial z} = 0. \tag{1}$$

Equation (1) is supplemented with the ideal gas equation of state

$$p = (\gamma - 1)\rho \left[e - \frac{1}{2}\left(v_x^2 + v_y^2 + v_z^2\right) \right].$$

The vector of conservative variables Q and the flux vectors F_x, F_y, F_z have the form

$$Q = \begin{pmatrix} \rho \\ \rho v_x \\ \rho v_y \\ \rho v_z \\ \rho e \end{pmatrix},$$

$$F_x = \begin{pmatrix} \rho v_x \\ \rho v_x v_x + p - \tau_{xx} \\ \rho v_x v_y - \tau_{xy} \\ \rho v_x v_z - \tau_{xz} \\ (\rho e + p)v_x - v_x \tau_{xx} - v_y \tau_{xy} - v_z \tau_{xz} + q_x \end{pmatrix},$$

$$F_y = \begin{pmatrix} \rho v_y \\ \rho v_y v_x - \tau_{yx} \\ \rho v_y v_y + p - \tau_{yy} \\ \rho v_y v_z - \tau_{yz} \\ (\rho e + p)v_y - v_x \tau_{yx} - v_y \tau_{yy} - v_z \tau_{yz} + q_y \end{pmatrix},$$

$$F_z = \begin{pmatrix} \rho v_z \\ \rho v_z v_x - \tau_{zx} \\ \rho v_z v_y - \tau_{zy} \\ \rho v_z v_z + p - \tau_{zz} \\ (\rho e + p)v_z - v_x \tau_{zx} - v_y \tau_{zy} - v_z \tau_{zz} + q_z \end{pmatrix}.$$

The components of the viscous stress tensor and the heat flux components are given by

$$\tau_{ij} = \mu \left(\frac{\partial v_i}{\partial x_j} + \frac{\partial v_j}{\partial x_i} - \frac{2}{3} \frac{\partial v_k}{\partial x_k} \delta_{ij} \right), \quad q_i = -\lambda \frac{\partial T}{\partial x_i}.$$

Here, t is the time, ρ is the density, v_x, v_y, v_z are the velocity components in the x, y and z directions, p is the pressure, e is the total energy per unit mass, T is the temperature, γ is the ratio of specific heat capacities.

The molecular viscosity, μ, is a function of temperature. It is modelled with Sutherland's law

$$\frac{\mu}{\mu_*} = \left(\frac{T}{T_*}\right)^{3/2} \frac{T_* + S_0}{T + S_0},$$

where μ_* and T_* are a reference viscosity and temperature, and S_0 is a constant determined experimentally, so that $\mu_* = 1.7894 \times 10^{-5}$ kg/(m·s), $T_* = 273.11$ K, and $S_0 = 110.56$ K for air. The thermal conductivity, λ, is linked to the specific heat capacity at constant pressure, c_p, and the Prandtl number, Pr, so that $\lambda = c_p \mu/\text{Pr}$, and $\text{Pr} = 0.7$ for air.

2.2. Euler Equations

Inviscid flow analysis neglects the effect of viscosity on the flow and is appropriate for high-Reynolds-number applications where inertial forces tend to dominate viscous forces. In particular, inviscid flow calculations are appropriate in an aerodynamic analysis of some high-speed projectile providing a good initial solution for problems involving complicated flow physics and flow geometry [1]. The viscous forces are still important, but in the early stages of calculations the viscous terms in the momentum equations are ignored. Once the calculation has been started and the residuals are decreasing, the viscous terms may be turned on (by enabling a laminar or turbulent flow) and the solution can be continued to convergence.

The unsteady flow of an inviscid compressible gas is described by Euler equations with a source term that takes into account energy supply. To describe flows with gas dynamic discontinuities appearing in high-speed flows, the integral form of the Euler equations is used

$$\frac{\partial}{\partial t} \int_V \boldsymbol{U}\, dV + \oint_S \boldsymbol{F}\, dS = \int_V \boldsymbol{H}\, dV. \qquad (2)$$

The vector of conservative variables, the flux vector and the source term have the form

$$\boldsymbol{U} = \begin{pmatrix} \rho \\ \rho \boldsymbol{v} \\ e \end{pmatrix}, \quad \boldsymbol{F} = \begin{pmatrix} \rho v_n \\ \rho v_n \boldsymbol{v} + p\boldsymbol{n} \\ (e+p)v_n \end{pmatrix}, \quad \boldsymbol{H} = \begin{pmatrix} 0 \\ 0 \\ q \end{pmatrix}.$$

The specific total energy is found as

$$e = \frac{p}{\gamma - 1} + \frac{1}{2}\rho v^2.$$

Here, t is the time, ρ is the density, v is the velocity, p is the pressure, e is the total energy per unit mass, n is the external unit normal to the boundary, v_n is the normal velocity ($v_n = v \cdot n = un_x + vn_y$), u and v are the Cartesian velocities in x and y coordinate directions, n_x and n_y are the projections of unit normal on x and y directions, γ is the ratio of specific heat capacities, and q is the specific power of energy supply. The intensity of energy supply is specified or found from additional relations taking into account physics of the specific problem. The energy of chemical transformation is the internal energy of the system and is taken into account as an integral part of the total energy. The rate of a chemical reaction is written in the Arrhenius form, and the pre-exponential factor and activation energy are specified for each specific chemical system.

For the model of a stationary one-dimensional flow with energy supply in a narrow zone (reaction front), there is an exact solution linking the flow parameters in the zones before and behind the energy supply. The relationship between the thermodynamic variables before and behind the energy supply zone is described by the Rankin–Hugoniot equation. The pressure and density behind the reaction front are related to the inlet Mach number by the Raylcigh–Michelson equation. The state of the medium behind the energy supply zone is determined from the intersection of the Michelson line with the Rankine–Hugoniot equation. At the same time, there is a maximum power supply that can be brought to the flow under steady-state conditions.

2.3. Initial and Boundary Conditions

To solve the problem in a stationary formulation, the pseudo-time method is used, according to which the steady state (time independent) distributions of gas parameters are found as a solution of the unsteady problem for $t \to \infty$. In this case, the form of the initial distribution of the flow quantities is insignificant. To speed up the calculation, physical initial conditions are used, based on the relations of the one-dimensional theory of isentropic gas flow in a nozzle. When solving the problem in an unsteady formulation, the steady state solution of the problem is selected for the initial approximation in the absence of energy supply. The solution of an unsteady problem when applying energy according to a periodic law is carried out until a periodic solution is obtained. The moment of achievement of a periodic solution is determined by comparing average values of the gas flow over a period of time that is a multiple of the period of energy supply (usually after 10 periods).

The no-penetration boundary condition for normal velocity is applied to the channel wall (the boundary condition for the tangential velocity on the wall is not set in the inviscid gas model). In the inlet section of the channel, the total pressure and total temperature are specified. In the case of a subsonic outflow from the nozzle, a static pressure equal to the surrounding pressure is fixed in the outlet section. This system of conditions determines the physical boundary conditions. Missing boundary conditions are determined from equations written in the characteristic form.

Negligible viscosity can no longer be assumed near solid boundaries. Assuming inviscid flow can be a useful tool in solving many fluid dynamics problems, however, this assumption requires careful consideration of the fluid sub-layers when solid boundaries are involved [12]. The conventional method for the analysis of supersonic nozzle flowfields is to assume the flow to be inviscid everywhere except near the wall where a thin viscous layer grows. The inviscid core is described by Euler equations and the wall layer is described by boundary layer equations, or Navier–Stokes equations are applied to flow domain. This technique is universally used to design contoured supersonic nozzles for specified exit flow conditions [13].

However, in supersonic flow, shock and boundary layer interaction is evident, and the structure of this interaction is complex and difficult to predict. Recent investigations in the inviscid and viscous interaction as well as more complex Navier–Stokes codes are encouraging, but still the supersonic flowfields with strong imbedded shock waves and boundary layer separations create tremendous difficulties. In the vicinity of the nozzle exit, the boundary layer at the initial jet expansion is very thin, and inviscid theory has been shown to describe the resulting jet flow reasonably well. Further from the nozzle exit, the thickening shear layer and recirculation region have required the addition of boundary layer approximations to inviscid theory to capture the jet structure fully. The effectiveness of all of these flow solution approaches are highly dependent on specific models applied and how certain models are coupled to the flow solver.

3. NUMERICAL METHOD

Two stages of solving the problem at one time step are distinguished: the energy stage (solving the equation of energy change together with the equations describing the heat generation in the flow due to external energy sources) and the gas dynamic stage (calculation of the density, velocity and pressure fields).

Using the finite volume method, the equation (2) is written in the form

$$U_i^{n+1} = U_i^n - \frac{\Delta t}{V_i} \sum_{j=1}^{N_e} F_{i,j} S_{i,j} + \Delta t H_i. \qquad (3)$$

Here, U_i is the vector of conservative variables, $F_{i,j}$ is the flux vector to the cell i through the edge j, $S_{i,j}$ is the edge length j of the cell i, V_i is the face area j of the cell i.

The Godunov method based on the solution of the Riemann problem is applied. When applying the finite volume scheme (3), the fluxes are calculated in the direction of the normal to the boundary. The flux on the edge of the control volume is determined as

$$F_{j+1/2} = \frac{1}{2}\big[F(U_L) + F(U_R)\big] - \frac{1}{2}|A|(U_R - U_L), \qquad (4)$$

where $|A| = R|\Lambda|L$, and $\Lambda = \mathrm{diag}\{v_n - a, v_n, v_n + a\}$ is the diagonal matrix of eigenvalues of the matrix A. For the ideal gas, the Jacobian is represented as

$$A = \begin{pmatrix} 0 & 1 & 0 \\ -(3-\gamma)\dfrac{u^2}{2} & (3-\gamma)u & \gamma - 1 \\ (\gamma/2 - 1)u^3 - \dfrac{ua^2}{\gamma - 1} & (3/2 - \gamma)u^2 + \dfrac{a^2}{\gamma - 1} & \gamma u \end{pmatrix}.$$

Matrices of the right and left Jacobian eigenvectors have the form

$$R = \begin{pmatrix} 1 & 1 & 1 \\ u - a & u & u + a \\ H - ua & u^2/2 & H + ua \end{pmatrix}, \quad L = \frac{1}{2}\begin{pmatrix} b_1 + u/a & -b_2 u - 1/a & b_2 \\ 2 - 2b_1 & 2b_2 u & -2b_2 \\ b_1 - u/a & -b_2 u + 1/a & b_2 \end{pmatrix}.$$

Here, $a = (\gamma p/\rho)^{1/2}$ is the speed of sound, and $b_1 = b_2 u^2/2$, $b_2 = (\gamma - 1)/a^2$.

To increase the order of accuracy of the finite volume scheme (4) when interpolating flow quantities on the edge of the control volume, the principle of minimal derivatives is used. The components of the vector of conservative variables on the faces of the control volume are determined by the Roe method. To avoid computational oscillations near the sound point, entropy correction is used. Time discretization is performed using the three-step Runge–Kutta method.

In the computations of inviscid flows some non-physical solutions such as expansion shocks may occur. The non-physical expansion shocks only occur in those regions of the computational domain where expansions are observed through sonic regions. Sonic expansion corresponds to the regions where the wave speed vanishes. Once the region of sonic expansion is detected, an expansion shock can be avoided by diffusing the expansion shock into the domain of computation within the band epsilon. The diffusion process is accomplished numerically by moving eigenvalues of the Jacobian away from its origin. Various formulations could diffuse the expansion shock.

Most high-order techniques experience a loss of robustness when the solution contains discontinuities or even under-resolved physical features. In order to avoid unrealistic solutions like expansion shocks from appearing as a part of a solution, the entropy condition for the Roe scheme must be satisfied. A variety of entropy fix formulae for the Roe scheme have been addressed in the literature [14, 15]. The most popular is the correction proposed in [16].

In cases where stationary energy supply is carried out in a supersonic flow (the energy supply intensity does not exceed a critical value), it is possible to design time-marching calculation schemes [17] (in a supersonic flow, the equations are hyperbolic in the x direction). To speed up the calculations, a vectorized approach to the fluxes calculation is used [18], which allows avoiding the programming of cyclic constructions.

4. REAL GAS EFFECTS

At low temperatures and pressures, internal energy consists of the energy of the translational and rotational motion of molecules. Therefore, air is considered as a perfect diatomic gas with a constant molecular weight, constant specific heats and constant adiabatic index. High-temperature flows are characterized by the excitation of vibrational degrees of freedom of polyatomic molecules, the presence of dissociation and ionization processes [1]. In the air, oxygen dissociation occurs at T =2000–4000 K, and nitrogen dissociation occurs at T =4000–10000 K. At T =7000–10000 K, the ionization process begins with the formation of free electrons. At T >10000–12000 K, significant fractions are single ions of these components.

To describe the equilibrium dissociation of some diatomic gases, the model of the ideal dissociating gas (Lighthill model) is used. Its properties are described by three constants, which allows for ja generalized analysis of the influence of dissociation. In the model developed in [19], air is considered as an ideal mixture of oxygen and nitrogen (it is assumed that there are no nitrogen compounds with oxygen) with constant molar concentrations, taking into account the excitation of vibrational and rotational degrees of freedom of molecules. The average molar mass of the mixture remains constant, and the equation of state retains the form corresponding to the equation of state of an ideal gas.

The advantage of the model proposed in [20] (the Kraiko model) is the inclusion of dissociation and ionization of air at high temperatures. When taking into account the real thermodynamic properties of air, explicit expressions are used for the density and specific internal energy through pressure and temperature $\rho = \rho(p, T)$ and $\varepsilon = \varepsilon(p, T)$. In the temperature range from 200 to 20000 K and pressures from 0.001 to 1000 atm, the model error does not exceed 1.5% in density and 3% in enthalpy.

The equations describing the flow of a real gas have the same form as the equations for an ideal gas (2). When using the approximate model (the Kraiko model or the Lighthill model), difficulties arise in the transition from conservative to physical variables. In approximate models, the gas state is determined in the function of the variables $\rho = \rho(p, T)$ and $\varepsilon = \varepsilon(p, T)$, and the dependence $p = p(\rho, \varepsilon)$ is used for calculations. For the transition between physical and conservative variables with known density and internal energy, a system of algebraic equations is solved

$$f_1(p, T) - \rho = 0, \quad f_2(p, T) - \varepsilon = 0.$$

In addition to the conservative form of equations, a quasi-linear form of equations is used in the calculations (for example, for specification of boundary conditions, determination of characteristic relations, calculation of eigenvectors). Unlike the perfect gas model, in the real gas model the Jacobian has the form

$$A = \begin{pmatrix} 0 & 1 & 0 \\ a^2 - u^2 - (H - u^2)\,p_\varepsilon/\rho & 2u - up_\varepsilon/\rho & p_\varepsilon/\rho \\ u(a^2 - H) - u(H - u^2)p_\varepsilon/\rho & H - u^2 p_\varepsilon/\rho & u + up_\varepsilon/\rho \end{pmatrix}.$$

For a gas whose thermal equation of state is written in the form $p = p(\rho, \varepsilon)$, the matrices composed of the right and left Jacobian eigenvectors have the form [21, 22]

$$R = \begin{pmatrix} 1 & 1 & 1 \\ u - a & u & u + a \\ H - ua & H - \rho a^2/p_\varepsilon & H + ua \end{pmatrix},$$

$$L = \frac{p_\varepsilon}{2\rho a^2} \begin{pmatrix} u^2 - H + \rho c(u + a)/p_\varepsilon & -u - \rho a/p_\varepsilon & 1 \\ 2(H - u^2) & 2u & -2 \\ u^2 - H - \rho c(u - a)/p_\varepsilon & -u + \rho a/p_\varepsilon & 1 \end{pmatrix}.$$

The total energy and local speed of sound are found as

$$H = \varepsilon + \frac{p}{\rho} + \frac{1}{2}u^2, \quad a^2 = p_\rho + p_\varepsilon \frac{p}{\rho^2},$$

where $p_\rho = \partial p/\partial \rho$, $p_\varepsilon = \partial p/\partial \varepsilon$. The Jacobian eigenvalues and characteristic ratios for real gas are preserved in the same form as in the case of an ideal gas flow. The vectors of conservative variables on the faces of the control volume are determined by the Roe method in the same relationships as for an ideal gas.

Methods for approximating thermodynamic functions are based on the use of an effective adiabatic index. The adiabatic exponent is replaced by a certain constant value, which is formally considered as an adiabatic exponent and corresponds to it with unexcited degrees of freedom of gas molecules. There are various approaches to determining the effective adiabatic exponent. In modelling shock wave processes, the definition of

$\gamma_s = h/\varepsilon$ is used. Using pressure and density as thermodynamic parameters, the internal energy is represented in the form $\varepsilon = p/[(\gamma_s - 1)\rho]$. The isentropic index (effective specific heat ratio) is determined by the formula

$$\gamma_e = \frac{c_p}{c_v} \left[1 + \frac{p}{\mu} \left(\frac{\partial \mu}{\partial p} \right)_T \right]^{-1},$$

where μ is the molar mass of gas, c_p and c_v are specific heat capacities at constant pressure and constant volume. At a pressure-dependent molecular weight, the adiabatic exponent γ_e does not equal (like γ_s) the ratio of specific heat capacities in the corresponding process. The values $\gamma_c = c_p/c_v$, γ_s and γ_e depend on the pressure and temperature as shown in the Figure 1a–c. The distribution of effective heat ratio as a function of temperature at a fixed pressure equal to $p = 1$ atm shows the Figure 1d.

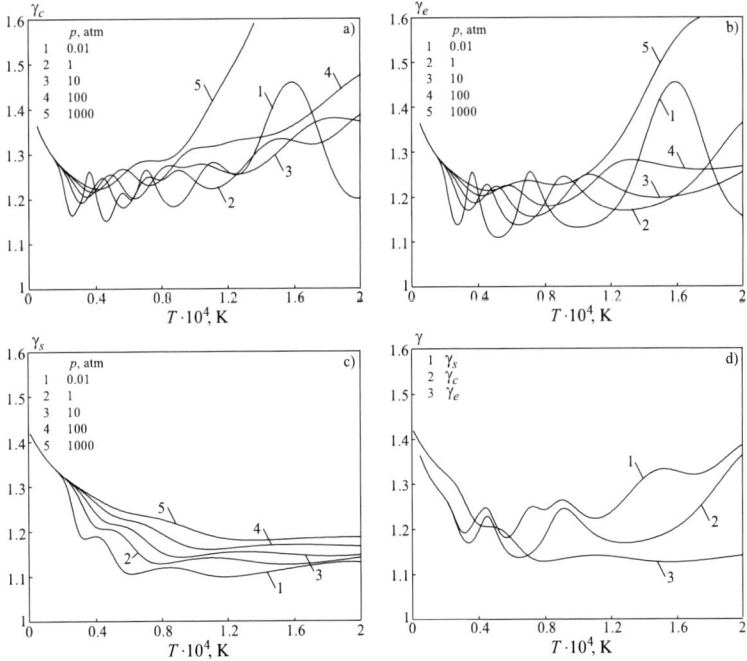

Figure 1. Dependencies of the effective ratio of specific heat capacities on temperature and pressure.

A solution of the Riemann problem for real gas is provided in [23]. The splitting of the flow vector and the increment of the flow vector for the case of a real gas are considered in [24, 25], and features of the Roe method are discussed in [21]. The case of a multi-dimensional flow is considered in [25], while various methods of splitting the flow vector are discussed in [26]. When using the Godunov method, it is assumed that the solution to the Riemann problem corresponds to the case of a frozen flow (for neighbouring cells, their effective values of the adiabatic exponents are used). The resulting frozen flows are used to calculate conservative variables, after which equilibrium parameters are determined for each cell.

Entropy and enthalpy are calculated with the equilibrium adiabatic index γ_e. Moreover, the equilibrium speed of sound is found from the relation $a_e = (\gamma_e p/\rho)^{1/2}$. The characteristic conditions are set similarly to the case of an ideal gas, taking $a^2 = p_\rho + p p_\varepsilon/\rho^2$.

5. MODEL OF ENERGY SUPPLY

In calculations of the flows of an ideal gas, simplified energy supply models are used which do not take into account processes of ionization and dissociation, as well as chemical reactions that occur in a real gas at high temperatures [10].

5.1. Temperature Distribution

The T-model assumes an instantaneous energy supply and zero value of the source term in the energy conservation equation ($q = 0$) [27]. The energy supply is specified by the temperature distribution at the initial time moment $t = 0$. The source term describes an ellipse centered at (x_0, y_0), and the dimensions of the half-axes R_x and R_y are determined as

$$T(x, y) = T_\infty + \Delta T(x, y),$$

where

$$\Delta T(x, y) = \Delta T_0 \exp\left[-\left(\frac{x - x_0}{kR_x}\right)^2 - \left(\frac{y - y_0}{kR_y}\right)^2\right].$$

Here, T_∞ is the temperature of surrounding gas, and $\Delta T_0 = T_0 - T_\infty$ is the temperature increment at the center of energy supply region. The smaller the parameter $0.5 \leqslant k \leqslant 1$, the smaller the gap in gas parameters at the boundary of the energy supply zone is observed. The value of the determining parameter of the model ΔT_0 is calculated by the formula

$$\Delta T_0 = \frac{Q}{\rho_\infty c_v I},$$

where Q is the energy absorbed by the medium, ρ_∞ is the density of surrounding gas, and c_v is the specific heat capacity at constant volume. The integral of the function describing the spatial distribution of intensity in the volume V_0 has the form

$$I = \int_{V_0} \exp\left[-\left(\frac{x-x_0}{kR_x}\right)^2 - \left(\frac{y-y_0}{kR_y}\right)^2\right] dV.$$

This model simulates only a single pulse of energy supply, not taking into account its duration.

5.2. Intensity Distribution

In q-model, it is possible to take into account the influence of stationary, single, pulse or periodic supply of energy of various durations and frequencies. In this case, the source term in the energy conservation equation is non-zero ($q \neq 0$) [27].

The specific power supplied to the region in the form of an ellipse centered at the point (x_0, y_0), and the dimensions of the semi-axes R_x and R_y are determined as

$$q(t, x, y) = q_0 f(t) \exp\left[-\left(\frac{x-x_0}{kR_x}\right)^2 - \left(\frac{y-y_0}{kR_y}\right)^2\right],$$

where q_0 is the specific power of the energy supply at the center, $f(t)$ is the function describing the intensity change in time, R_x and R_y are the characteristic lengths of the energy release region. The parameter $0.5 \leqslant k \leqslant 1$ determines the intensity values at the boundary of the energy supply region. For a single pulse of energy of duration τ function $f(t)$ has the form

$$f(t) = \begin{cases} 1 & \text{if } 0 \leqslant t < \tau, \\ 0 & \text{if } \tau \leqslant t. \end{cases}$$

The determining parameter of the model is calculated by the formula

$$q_0 = \frac{Q}{\rho_\infty \tau I}.$$

When the energy supply simulates the operation of a three-phase current plasma torch, in which the areas of energy release (electric arcs) periodically appear and move along the channel centreline, the coordinates $x_0 = x_0(t)$ and $y_0 = y_0(t)$ are determined by the path travelled by the arc and the speed.

In case of setting the mass energy supply, the effective energy supply is several times smaller due to a decrease in the gas density during heating [28]. Available results indicate the promise of using simplified models of energy supply for an ideal perfect gas during parametric calculations [10].

6. RESULTS AND DISCUSSION

Results of numerical simulation of one-dimensional and two-dimensional under- and over-expanded nozzle flows with a moving region of energy supply are presented.

6.1. Nozzle Geometry and Energy Supply

The channel cross-sectional area varies according to the dependence $S = 1 + 2x^2$, where $-0.3 \leqslant x \leqslant 1$. The coordinate $x = 0$ corresponds to the critical section of the nozzle. For one-dimensional calculations, a mesh containing 800 nodes is used. In two-dimensional calculations, a uniform mesh is used along the coordinates x and y containing 800×400 nodes. The number of mesh nodes is selected based on checking the convergence of numerical solutions on different meshes with a gradual increase in relevant dimension. Calculations are terminated upon reaching a final given point in time.

The energy supply model with a moving energy release region allows one to take into account the effect of the displacement of the intense energy supply zone in case of one or more plasma torch arcs. The diagram of a single energy supply region that periodically appears and moves in the channel is shown in the Figure 2a. The family of lines correspond to different points in time (with a constant step), and the shape of each curve displays the spatial distribution of the energy supply intensity. The energy supply diagram for three periodic moving arcs with a time-varying intensity is shown in the Figure 2b. In both cases, three full periods of energy supply are shown, where τ is the period of energy supply, and Q_0 is the maximum value of energy supply. The arc speed is 10 m/s, and the path travelled by the arc is 0.2 m. The arc burning time is 0.025 s.

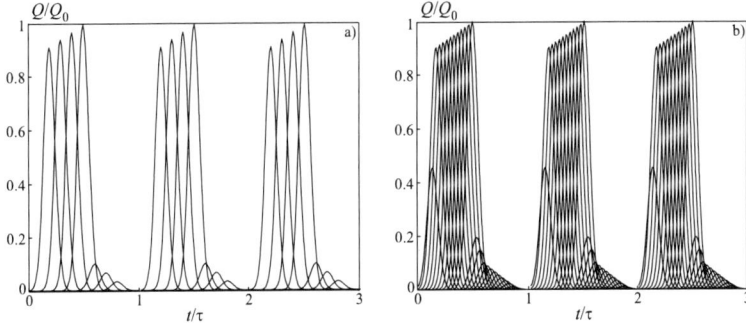

Figure 2. Distribution of the energy supply intensity over time for one arc (a) and three arcs (b).

The form of the function that describes the change in intensity over time is shown in the Figure 3 for a single arc and three arcs of a plasma torch. With a sufficiently large number of pulses, the distribution of the energy release intensity over the spatial coordinate becomes almost uniform.

The use of pulse or periodic energy sources makes it possible to realize sufficiently high peak power supply. The structure of the shock wave and the thermal wake behind the energy supply region substantially depend on the pulse repetition rate. With a certain frequency of such sources and by ensuring energy equivalence, it is possible to realize flows with properties close to those that are formed under conditions of stationary energy supply, when the thermal trace becomes continuous. The condition of energy equivalence is determined by the equality of the energy supply parameter in stationary and non-stationary cases.

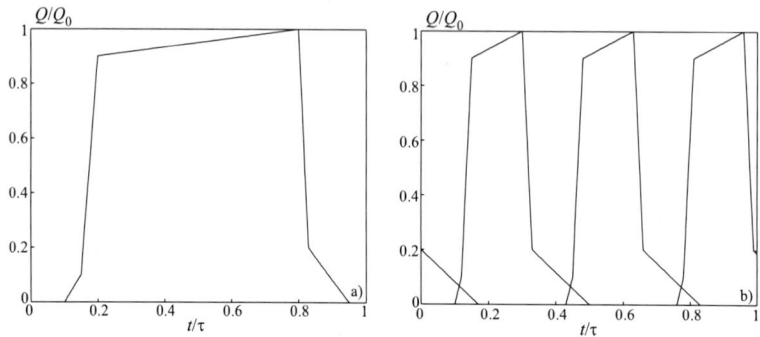

Figure 3. A single discharge pulse (a) and multiple discharge pulses (b).

6.2. Test Cases

The unsteady flows in a channel with a variable cross-sectional area in one-dimensional and two-dimensional formulations without an energy supply are considered, as well as a stationary one-dimensional flow with energy supply in a narrow zone (reaction front), which has an exact solution describing the flow quantities in zones before and behind power supply.

The pressure field in a two-dimensional flow in a nozzle without energy supply, obtained by the pseudo time-marching method, is shown in the Figure 4. Such flow regimes in the nozzle are considered when a nozzle shock wave is formed inside. The nozzle shock is visible in the region of concentration of the pressure level lines.

A comparison of the exact solution of the stationary quasi one-dimensional problem (solid line) and the limiting numerical solution of the unsteady two-dimensional problem (the symbols ∘ and ∗) is shown in the Figure 5. Satisfactory agreement of solutions is ensured on a relatively coarse mesh, and the intensity and position of the shock wave corresponds to the exact solution. The difference scheme spreads the nozzle shock wave over 1–2 mesh cells. The solution is monotonous (there are no non-physical oscillations).

Another test case corresponds to the flow in the channel, in the output section of which the boundary conditions are set corresponding to the supersonic flow (characteristic boundary conditions). The energy supply zone is reproduced in one cell. The pressure and temperature distributions

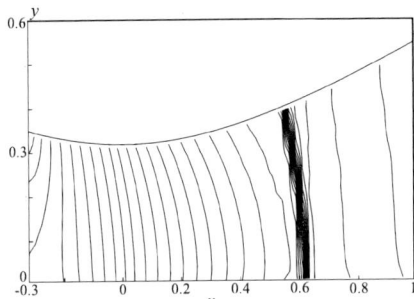

Figure 4. Contours of pressure in the nozzle flow without energy release.

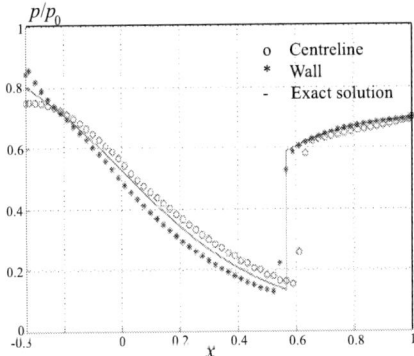

Figure 5. Comparison of exact solution (solid line) with results of numerical calculations (symbols o and *). Symbols o correspond to the flow quantities along the centreline, and symbols * correspond to the flow quantities near the nozzle wall.

behind the front of the power supply are shown in the Figure 6 depending on the Mach number in front of the power supply zone. For each Mach number, there are two solutions, and there is a threshold of energy supply intensity (critical energy supply). When the energy supply intensity exceeds a critical value, there is no stationary solution of the problem. For example, for the Mach number $M_1 = 3$, the critical energy supply is $q = Q/(c_p T_1) = 1.4815$. Subscripts 1 and 2 refer to flow quantities before and behind the energy supply zone.

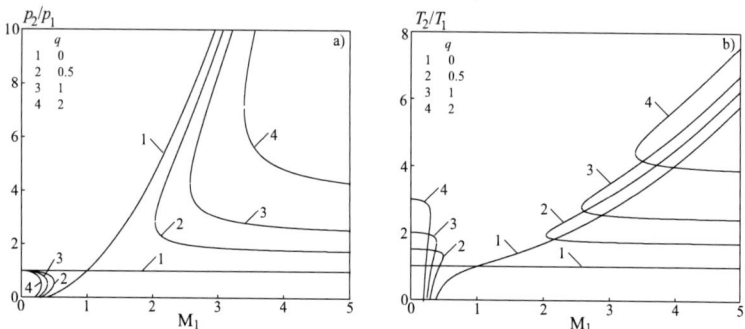

Figure 6. Pressure distributions (a) and temperature distributions (b) before and behind the energy supply region for $q = 0$ (1), 0.5 (2); 1 (3); 2 (4).

The distributions of the flow quantities during supercritical energy supply are shown in the Figure 7 at a fixed time, with $M_1 = 3$. The dimensionless energy supply parameter is assigned the value $q = 2$.

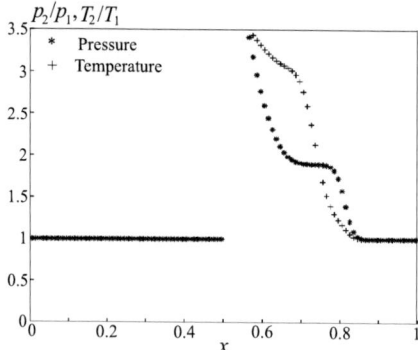

Figure 7. Pressure distributions (symbols $*$) and temperature distributions (symbols $+$) at time $t = 1.546$.

6.3. One-Dimensional Flows

The nozzle operating conditions are considered when the pressure of the over-expanded flow is restored through the nozzle shock wave. Unsteady energy supply, carried out in the section of the subsonic path of the nozzle, causes the development of shock waves accompanied not only by a change in the flow quantities, but also by the movement of shock wave structures and pressure fluctuations. Unsteady energy supply leads to a significant restructuring of the nozzle flow, an increase in temperature, and a change in gas pressure near the energy supply region. After the end of the pulse, the high-temperature zone is transferred along the centreline.

The pressure and temperature distributions at different points in time are shown in the Figure 8 and Figure 9. The total pressure and the total temperature in the inlet section are fixed at $8.5 \cdot 10^5$ Pa and 300 K, and the static pressure in the outlet section of the nozzle is set to $6 \cdot 10^5$ Pa. The solid lines show the pressure and temperature distributions corresponding to the stationary solution to the problem without energy supply, and the points show the pressure and temperature distributions in the unsteady case. The distributions of flow quantities given in the Figure 8a and Figure 9a correspond to the beginning of the arc ignition process. For simplicity, the position of the arc, the intensity of which has a Gaussian distribution in space with a sufficiently high localization, is shown by a vertical line for the initial period of the energy supply cycle. The energy supply causes the appearance of high gradients of flow quantities, localized in a narrow region. The process development is shown in the Figure 8b–d and Figure 9b–d. With a change in the intensity of energy supply, which corresponds to the ignition and extinction of an arc, the intensity of discontinuities changes, and regions with sharp gradients of the flow quantities move along the nozzle.

The effect of unsteady energy supply is shown in the Figure 10, where $q = Q_0/(G_0 H_0)$ is the dimensionless parameter that determines the energy supply intensity, and G_0 and H_0 are the mass flow rate and total enthalpy without energy supply. If there is a periodic energy supply in the subsonic part of the nozzle, the gas flow through the critical section also becomes a periodic function. The integral value of the flow rate decreases compared to its value in a flow without energy supply, and this difference increases with an increase in the energy supply intensity.

For alternating current plasmatrons, the energy supply option with sev-

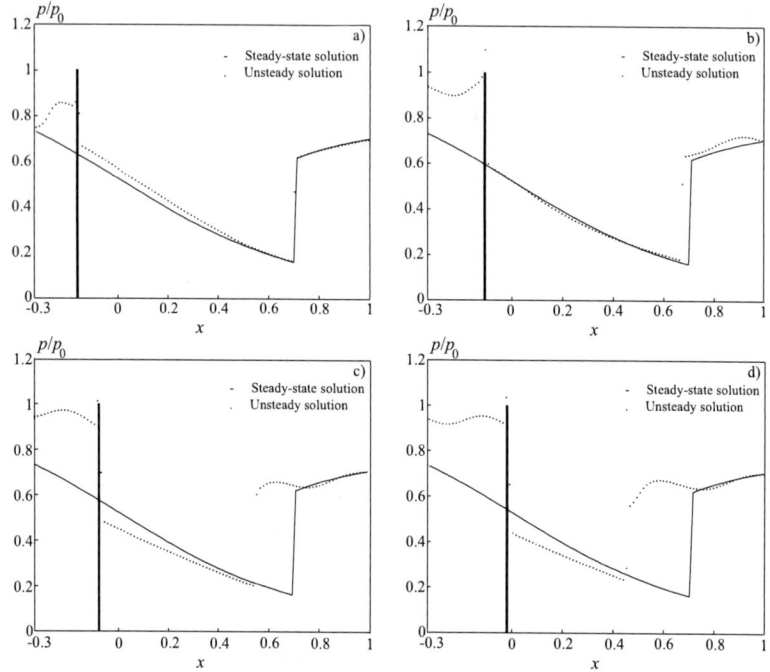

Figure 8. Pressure distributions at time $t/\tau = 1.156$ (a), 1.396 (b), 1.516 (c), 1.724 (d) for one arc.

eral simultaneously moving arcs is of interest. Figure 11 and Figure 12 show the effect of heat release on the pressure and temperature distributions for three arcs (these results correspond to one energy supply cycle). The arcs move one after the other along the nozzle centreline. The vertical lines correspond to the positions of the arc at the corresponding time moments. For simplicity, the spatial distribution of the energy supply intensity is not shown in the figures.

Due to the fact that the arcs move periodically, the distributions of flow quantities in any sections are also periodic functions (Figure 13). An increase in the number of arcs leads to a more uniform temperature distribution along the nozzle centreline.

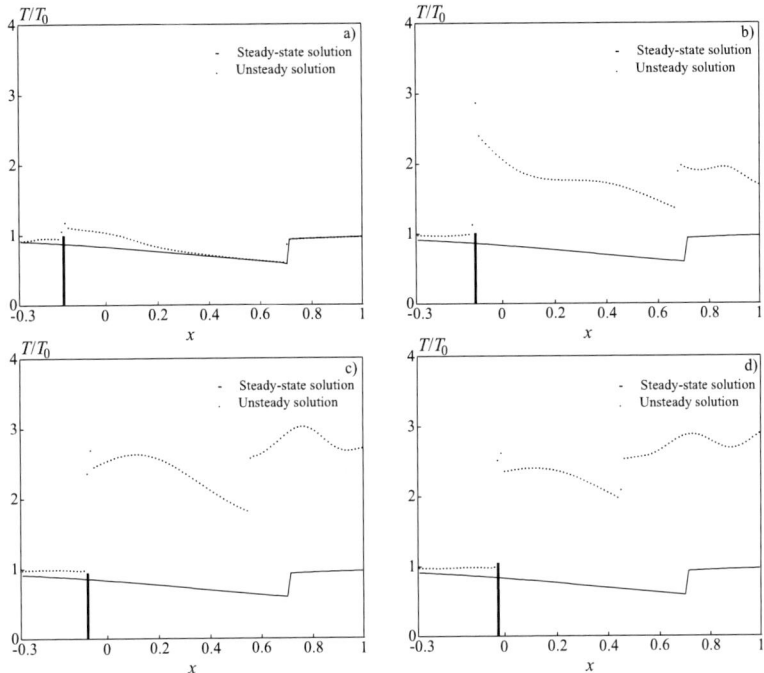

Figure 9. Temperature distributions at time $t/\tau = 1.156$ (a), 1.396 (b), 1.516 (c), 1.724 (d) for three arcs.

6.4. Two-Dimensional Flows

In the two-dimensional case, energy is supplied in the subsonic part of the nozzle. The energy supply region moves along the nozzle centreline, and the energy supply intensity cyclically changes in time. The case of one burning arc is considered.

The pressure and temperature distributions at different times of the energy supply cycle are shown in the Figure 14 and Figure 15. At moderate intensities of energy supply, its effect on the pressure field is manifested to a lesser extent than on other flow quantities. The origin of the high-temperature region shows the Figure 14a and Figure 15a, which correspond to the moment the arc begins to burn. The high-temperature region is transferred by the flow. The length of this region increases in time, and

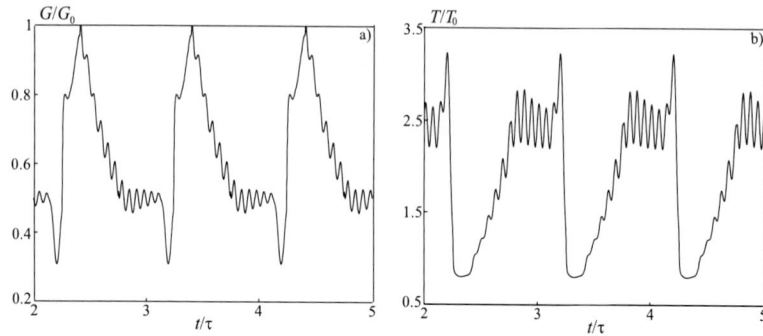

Figure 10. Mass flow rate through critical section of the nozzle (a) and temperature distribution in the critical section (b) for one arc.

temperature gradients increase as well (Figure 14b and Figure 15b). Behind the shock wave, a trace of the previous heat spot is visible. Further development of the high-temperature region is observed in Figure 14c and Figure 15c. After the arc attenuation, a convective drift of the heat spot occurs, and the temperature field becomes the same as during a flow without energy supply (Figure 14d and Figure 15d). The configuration of the perturbed region and the amplitude values of the flow quantities (temperature, pressure, Mach number) depend on the size of the energy release region and the maximum value of the energy input.

The effect of the energy supply on the distribution of flow quantities is shown in Figure 16 and Figure 17 for a fixed nozzle section. The solid lines correspond to the critical section of the nozzle, and the dash-dotted lines correspond to the outlet section of the nozzle. The influence of the energy supply on the temperature field is much more significant than that on the pressure field. The pressure distributions inside and outside the energy supply region are similar. The results obtained allow concluding that there is a significant non-uniform flow in the outlet section of the nozzle. Both the temperature field and the pressure field have two areas of sharp changes in parameters. In addition to the nozzle shock wave, sharp gradients of flow quantities are observed in a narrow region of intense energy release.

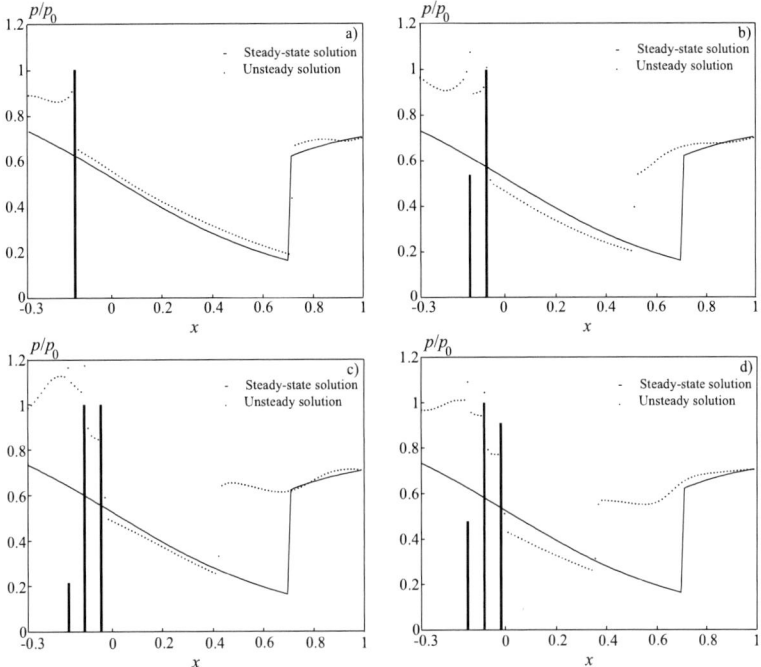

Figure 11. Pressure distributions at time $t/\tau = 1.224$ (a), 1.544 (b), 1.668 (c), 1.768 (d) for three arcs.

6.5. Flows of Real Gas

The nozzle cross sectional area changes as $S(x) = 1 + 2x^2$, where $-1 \leqslant x \leqslant 2$. The total temperature at the inlet section of the nozzle is fixed at 5000 K. At the nozzle exit, a pressure is set corresponding to the case of subsonic outflow. The calculations are carried out using various methods for calculating the effective adiabatic exponent (γ_c, γ_e, γ_*). The results are compared with those based on ideal gas model for two values of the adiabatic index ($\gamma_1 = 1.4$ and $\gamma_2 = 1.2$). The influence of various methods for determining the effective adiabatic index on the distribution of flow quantities is investigated.

Comparative distributions of pressure, temperature and Mach number for various methods for determining the effective adiabatic exponent are

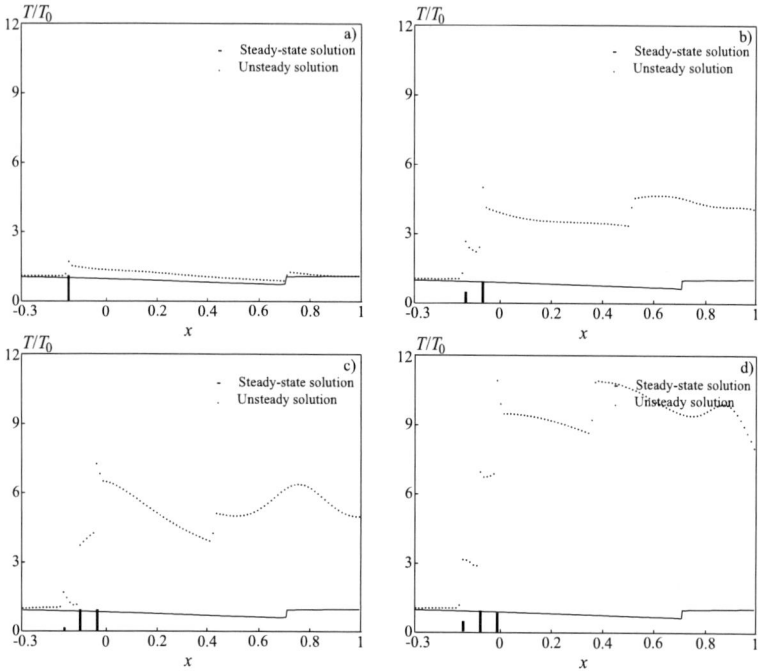

Figure 12. Temperature distributions at time $t/\tau = 1.224$ (a), 1.544 (b), 1.668 (c), 1.768 (d) for three arcs.

shown in the Figure 18 for a flow without an internal shock wave. The solid and dash-dotted lines correspond to the adiabatic exponents $\gamma_1 = 1.4$ and $\gamma_2 = 1.2$, and the symbols ○, ∗ and + correspond to the adiabatic exponents γ_c, γ_e and γ_s. In the calculations corresponding to ideal and real gas flows, equal values of temperature and pressure are set at the nozzle inlet. Moreover, the total enthalpy of real gas is much higher than for an ideal gas.

Comparative distributions of pressure, temperature, and Mach number are shown in the Figure 19 during the flow through the nozzle with the formation of an internal shock wave (the notation is the same as in the Figure 18). The change in various effective adiabatic exponents along the nozzle centreline is shown in the Figure 20 (the solid line corresponds to γ_c, the dashed line corresponds to γ_s, the dash-dotted line corresponds to γ_e).

Simulation of High-Temperature Flows in Nozzles ... 51

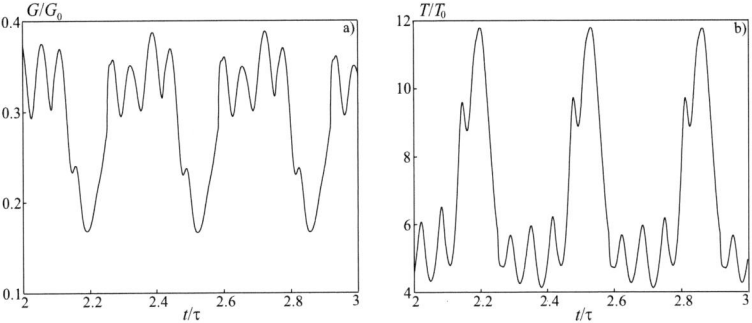

Figure 13. Mass flow rate through the critical section of nozzle (a) and temperature distribution in the critical section of nozzle (b) for three arcs.

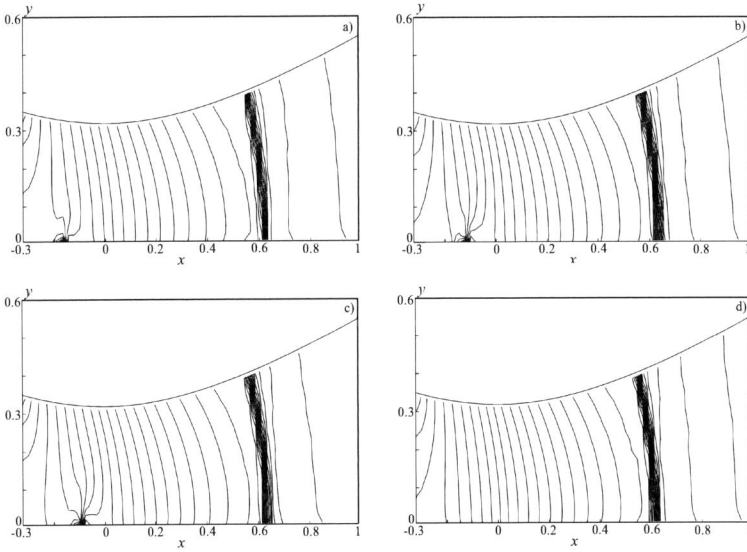

Figure 14. Contours of pressure at time $t/\tau = 2.154$ (a), 2.410 (b), 2.531 (c), 2.977 (d).

The pressure distribution in the nozzle flow with a subsonic output depends both on the magnitude of the adiabatic index and method for determining

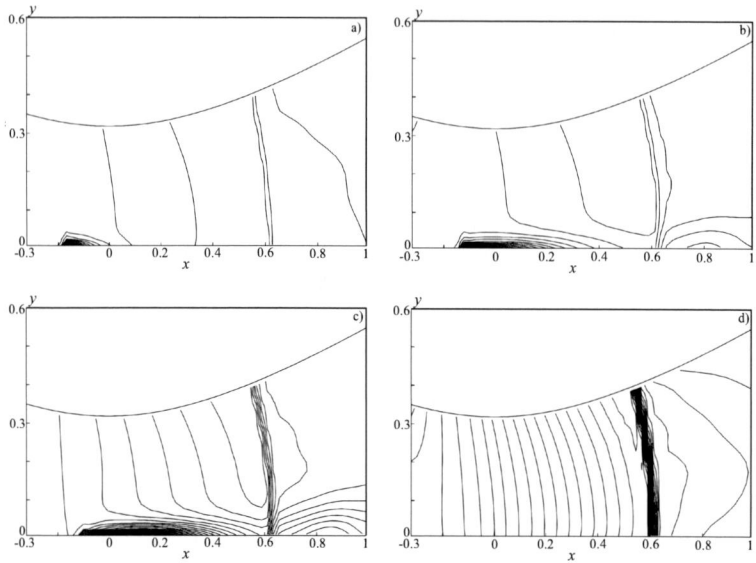

Figure 15. Contours of temperature at time $t/\tau = 2.154$ (a), 2.410 (b), 2.531 (c), 2.977 (d).

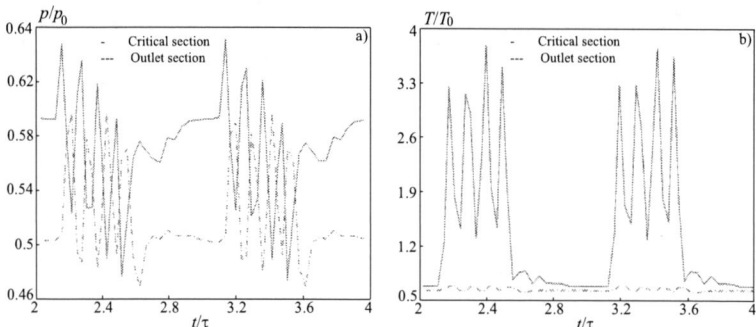

Figure 16. Pressure (a) and temperature (b) distributions along nozzle centreline ($x = 0$).

its effective value. The distribution of the Mach number is more sensitive

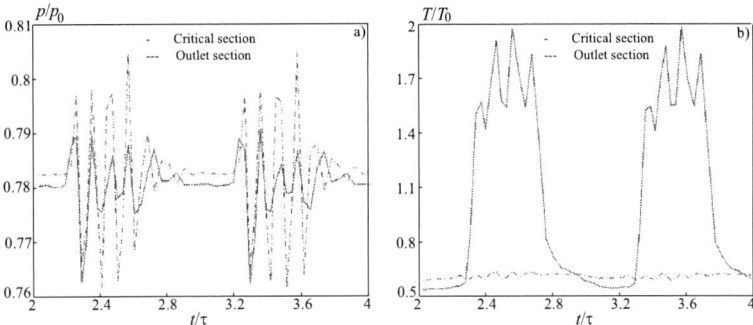

Figure 17. Pressure (a) and temperature (b) distributions near nozzle wall ($x = 0.98$).

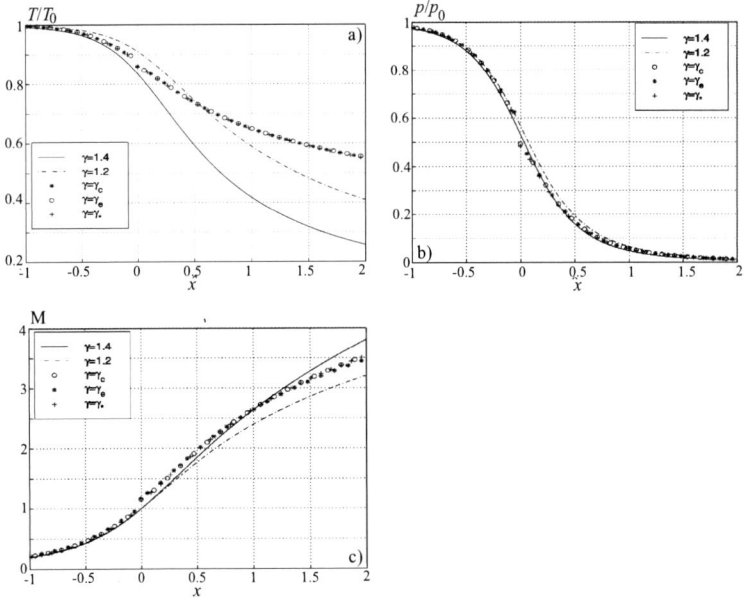

Figure 18. Temperature (a), pressure (b) and Mach number (c) distributions for various effective specific heat ratios for under-expanded nozzle flow.

to the method of determining the adiabatic exponent. Using a real gas model leads to significant deviations in the temperature distribution. Due to the significantly different dissociation energies for oxygen and nitrogen, oxygen begins to dissociate earlier than nitrogen. Behind the shock wave, in accordance with an increase in temperature, the degree of dissociation also rises sharply.

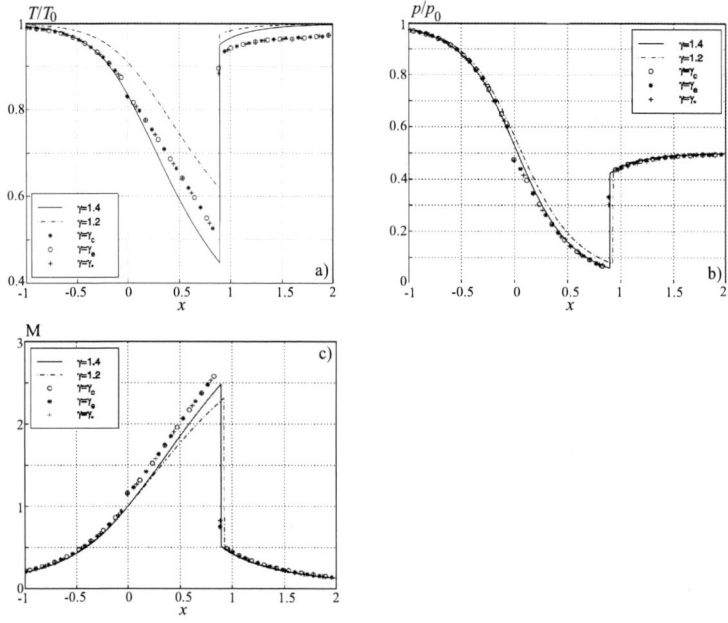

Figure 19. Temperature (a), pressure (b) and Mach number (c) distributions for various effective specific heat ratios for the over-expanded nozzle flow.

Another option for comparing ideal and real gas flows, corresponding to the case of an under-expanded flow in the nozzle, is shown in the Figure 21a. The same total pressure and total enthalpies are taken as conditions at the nozzle inlet. The temperature distribution in real gas is much lower than in ideal gas. This is because real gas absorbs significant energy for the dissociation of molecules. For another calculation option, corresponding to the case of an over-expanded flow in the nozzle, the temperature is assumed to be 10000 K (Figure 21b). For such conditions, ionization

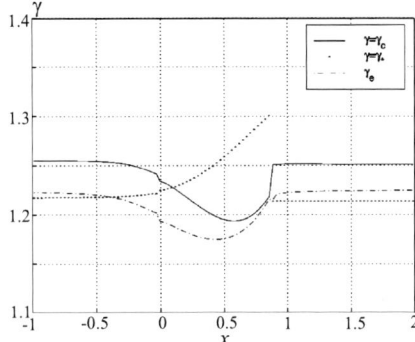

Figure 20. Distributions of effective specific heat ratios along the nozzle centreline.

processes in the input section become noticeable.

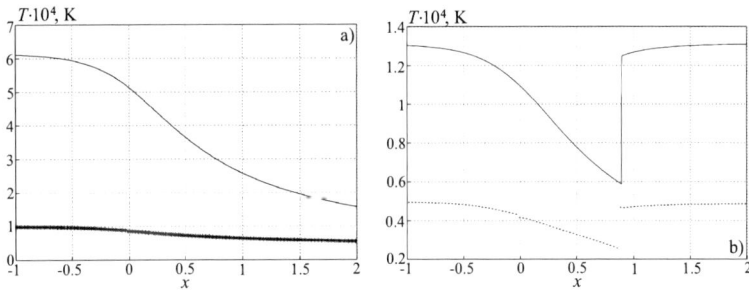

Figure 21. Temperature distributions of ideal gas and real gas for under-expanded (a) and over-expanded (b) nozzle flows.

A real gas model allows taking into account dissociation processes. The proportion of dissociated oxygen and nitrogen molecules depends on the current temperature and pressure. It should be noted that, due to the significantly different dissociation energies for oxygen (5.1 eV) and nitrogen (9.8 eV), oxygen begins to dissociate earlier than nitrogen. By the time nitrogen dissociation begins, oxygen dissociation is almost complete (Figure 22a). The ionization process is just starting, and the dissociation of oxygen completely finishes to the start of ionization. The change in the degree of dissociation of oxygen and nitrogen molecules in the case of over-expanded outflow from the nozzle is shown in the Figure 22b. Behind the shock wave, in accordance with an increase in temperature, the degree of dissociation also rises sharply.

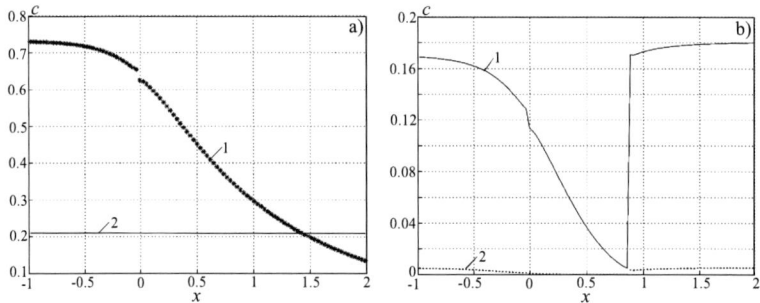

Figure 22. Distributions of molar concentrations of dissociated oxygen molecules (line 1) and nitrogen (line 2) for under-expanded (a) and over-expanded (b) flows.

CONCLUSION

Numerical studies of gas flows in technical devices in which processes associated with unsteady supply of energy are carried out. To calculate one-dimensional and two-dimensional nozzle flows with unsteady energy supply in the subsonic part of the nozzle, the finite volume method was applied in the framework of ideal and real gas models.

The operating conditions of the nozzle with significant over-expansion are considered, when the pressure of the over-expanded flow is restored through the nozzle shock wave. The dependencies of the flow quantities on the intensity of energy supply and its effect on the position and intensity of the nozzle shock wave are reported. The supply of thermal energy to the flow leads to a significant increase in temperature and a change in gas pressure near the energy supply region. After the end of the pulse, the zone of high temperature is transferred along the nozzle centreline.

Unsteady energy supply, carried out in the subsonic part of the nozzle, causes the development of intense shock wave processes, accompanied not only by a change in the flow quantities in the nozzle, but also by the movement of shock waves and pressure oscillations. The result of energy release in the flow is an increase in temperature and pressure, but a significant decrease in the Mach number downstream. In the presence of a periodic energy supply, the mass flow rate through the critical section becomes a periodic function. The integral value of the flow rate decreases compared to its value in a flow without an energy supply, and with an increase in the energy supply intensity, this difference increases. At moderate intensities of energy supply, its effect on the pressure field is manifested to a lesser extent than on other flow quantities. The configuration of the perturbed region and the amplitudes of the flow quantities (pressure, temperature, Mach number) depend on the size of the energy release region and the maximum energy input.

ACKNOWLEDGMENT

The research was supported by Russian Science Foundation No. 21-19-00657, https://rscf.ru/en/project/21-19-00657/.

REFERENCES

[1] Anderson, J.D. (2006). Hypersonic and high temperature gas dynamics. *AIAA*.

[2] Zamuraev, V.P. and Kalinina, A.P. (2010). Gas-dynamic effects of periodic energy input into a divergent channel. *Technical Physics*, 55(1), 40–43.

[3] Zabaikin, V.A., Naumov, I.E. and Tretyakov, P.K. (2012). Change in the regimes of flow and combustion in a channel under external energy action. *Journal of Engineering Physics and Thermophysics*, 85(6), 1331–1338.

[4] Chikitkin, A., Petrov, M., Dushkov, R. and Shifrin, E. (2018). Aerodynamic design of a Laval nozzle for real gas using hodograph method. *Aerospace*, 5, 96–112.

[5] Latypov, A.F. (2009). Numerical simulation of the flow in a variable-section channel with pulsed-periodic energy supply. *Applied Mechanics and Technical Physics*, 50(1), 3–11.

[6] Fang, M.T.C., Zhuang, Q. and Shen, M.Y. (1994). The computation of axisymmetric, supersonic nozzle arc using adaptive grids. *IEEE Transactions on Plasma Science*, 22(3), 228–234.

[7] Yan, J.D., Fang, M.T.C. and Jones, C. (1997). Electrical and aerodynamic behaviour of arcs under shock conditions. *IEEE Transactions on Plasma Science*, 25(5), 840–845.

[8] Surzhikov, S.T. (1995). Mathematical models of subsonic Laval nozzles of laser-plasma accelerators. *High Temperature*, 33(3), 435–448.

[9] Yegorov, I.V. and Ivanov, D.V. (1997) Simulation of the flow with non-equilibrium chemical reactions in a channel of various cross section. *Mathematical Modelling*, 9(11), 85–100.

[10] Zheltovodov, A.A. and Pimonov, E.A. (2013). Numerical simulation of an energy deposition zone in quiescent air and in a supersonic flow under the conditions of interaction with a normal shock. *Technical Physics*, 83(2), 170–184.

[11] Brykov, N.A., Emelyanov, V.N., Karpenko, A.G. and Volkov, K.N. (2021). Flows of real gas in nozzles with unsteady local energy supply. *Computers and Mathematics with Applications*, 81, 702–724.

[12] Kubo, K., Miyazato, Y. and Matsuo, K. (2010). Study of choked flows through a convergent nozzle. *Journal of Thermal Science*, 19(3), 193–197.

[13] Bakhtian, N.M. and Aftosmis, M.J. (2011). Analysis of inviscid simulations for the study of supersonic retropropulsion. *AIAA Paper*, 2011-3194.

[14] Pelanti, M., Quartapelle, L. and Vigevano, L. (2011). *A review of entropy fixes as applied to Roe's linearization*. Aerospace and Aeronautics Department of Politecnico di Milano.

[15] Carpenter, M.H. (2013). High-order entropy stable formulations for computational fluid dynamics. *AIAA Paper*, 2013-2868.

[16] Harten, A. and Hyman, J.M. (1983). Self-adjusting grid methods for one-dimensional hyperbolic conservation laws. *Journal of Computational Physics*, 50, 235–269.

[17] Emelyanov, V.N., Pustovalov, A.V. and Volkov, K.N. (2019). Supersonic jet and nozzle flows in uniform-flow and free-vortex aerodynamic windows of gas lasers. *Acta Astronautica*, 163, 232–243.

[18] Volkov, K.N. (2018). Multigrid and preconditioning techniques in CFD applications. *CFD Techniques and Thermo-Mechanics Applications*, 83–149.

[19] Levin, V.A., Gromov, V.G. and Afonina, N.E. (2000). Numerical analysis of the effect of local energy supply on the aerodynamic drag and heat transfer of a spherically blunted body in a supersonic air flow. *Applied Mechanics and Technical Physics*, 41(5), 915–922.

[20] Kraiko, A.N. and Makarov, V.E. (1996). Explicit analytic formulas defining the equilibrium composition and thermodynamic functions of air for temperatures from 200 to 20000 K. *High Temperature*, 34(2), 202–213.

[21] Glaister, P. (1988). An approximate linearised Riemann solvers for the Euler equations for real gases. *Journal of Computational Physics*, 74(2), 382–408.

[22] Smenov, A.Yu. (1997). A modified Courant–Isaacson–Rees method for gas dynamics with an arbitrary equation of state. *Computational Mathematics and Mathematics Physics*, 37(11), 1334–1340.

[23] Colella, P. and Glaz, P.M. (1985). Efficient solution algorithms for the Riemann problem for real gases. *Journal of Computational Physics*, 59(2), 264–289.

[24] Vinocur, M. and Liu, Y. (1988). Equilibrium gas flow computations: an analysis of numerical formulations of conservation laws. *AIAA Paper*, 88-0127.

[25] Grossman, B. and Walters, R.W. (1989). Analysis of flux-split algorithms for Euler's equations with real gases. *AIAA Journal*, 27(5), 524–531.

[26] Liou, M.-S. and van Leer, B. (1990). Splitting of inviscid fluxes for real gases. *Journal of Computational Physics*, 87(1), 1–24.

[27] Yan, H., Adelgren, R., Boguszko, M., Elliott, G. and Knight, D. (2003). Laser energy deposition in quiescent air. *AIAA Journal*, 41(10), 1988–1995.

[28] Kucherov A.N. (2009). Some problems of gas flows with definite distributed heat sources. *TsAGI Scientific Notes*, 40(4), 3–14.

In: Navier-Stokes Equations ...
Editor: Peter J. Johnson

ISBN: 978-1-53619-967-3
© 2021 Nova Science Publishers, Inc.

Chapter 3

INTEGRALS OF THE NAVIER – STOKES AND EULER EQUATIONS FOR MOTION OF INCOMPRESSIBLE MEDIUM

Alexander V. Koptev[*]

Math. Dept., Admiral Makarov State University
of Maritime and Inland Shipping, Saint-Petersburg, Russia

Abstract

We consider integrals of 3D motion of a viscous and ideal medium. A study of relationships between integrals and areas of their applicability has been carried out for motion of incompressible medium. As a result of applying the methods of partial differential equations and mathematical physics it is shown that all considered integrals could be unite with the chain as a tree. On the base of a tree located the first integral of 3D Navier – Stokes equations obtained by the author. This integral plays the role of a root integral. All other integrals united by the chain under consideration are its special cases. A proof is given that each of the well-known classical integrals of Bernoulli, Euler-Bernoulli and Lagrange-Cauchy is a special case of the root integral. So special cases of the root integral are six new ones obtained by the author and three well-known integrals of Lagrange-Cauchy, Bernoulli and Euler-Bernoulli.

---

[*]Corresponding Author's Email: Alex.Koptev@mail.ru.

Keywords: motion, incompressible medium, viscous fluid, ideal fluid, equation, velocity, pressure, integral, chain, tree

1. Introduction

For case of motion of an incompressible medium the density and all other physical characteristics are constant and main unknowns are the components of the velocity vector u, v, w and pressure p. The Navier — Stokes equations [1-2] are generally accepted to describe motion of such a medium. Subject to the existence of the potential of external forces in dimensionless variables they can be represented as

$$\frac{\partial u}{\partial t} + u\frac{\partial u}{\partial x} + v\frac{\partial u}{\partial y} + w\frac{\partial u}{\partial z} = -\frac{\partial(p+\Phi)}{\partial x} + \frac{1}{Re}\Delta u, \quad (1)$$

$$\frac{\partial v}{\partial t} + u\frac{\partial v}{\partial x} + v\frac{\partial v}{\partial y} + w\frac{\partial v}{\partial z} = -\frac{\partial(p+\Phi)}{\partial y} + \frac{1}{Re}\Delta v, \quad (2)$$

$$\frac{\partial w}{\partial t} + u\frac{\partial w}{\partial x} + v\frac{\partial w}{\partial y} + w\frac{\partial w}{\partial z} = -\frac{\partial(p+\Phi)}{\partial z} + \frac{1}{Re}\Delta w, \quad (3)$$

$$\frac{\partial u}{\partial x} + \frac{\partial v}{\partial y} + \frac{\partial w}{\partial z} = 0. \quad (4)$$

where Δ is a three-dimensional Laplace operator on spatial coordinates,

$$\Delta = \frac{\partial^2}{\partial x^2} + \frac{\partial^2}{\partial y^2} + \frac{\partial^2}{\partial z^2};$$

Φ is the potential of external forces which is a given function;
Re is the Reynolds number which is a dimensionless non-negative parameter

$$Re = \frac{U_0 L}{\nu}.$$

U_0 and L in the last formula denotes scales of velocity and length;
ν is the kinematic viscosity coefficient.

The vanishing of the kinematic viscosity corresponds to $Re \to +\infty$. For this case right-hand sides of (1-3) significantly simplified. They fall terms proportional to $\frac{1}{Re}$ and the Navier — Stokes equations pass into Euler equations. Term as "ideal" or "perfect" fluid are usually used for such a medium.

Navier — Stokes and Euler equations have numerous applications to practical problems. Traditional areas of applicability include shipbuilding, aircraft manufacturing, hydrometeorology, hydrology, tribology, cardiology. Many problems arising in these areas have been resolved precisely thanks to the Navier - Stokes and Euler equations. But these equations are of great interest in a purely mathematical sense. Their complex study is one of the areas of modern mathematical physics [3-4]. Today, however, many issues not fully clarified and require further study. One of the main problems is the lack of a constructive method for solution. An important step in this direction is the construction of integrals. The provisions of classical fluid mechanics in this regard are as follows. There are three well-known integrals convenient for practical application [1-2].

For potential non-stationary motion of incompressible medium occurs the Lagrange — Cauchy integral. For an incompressible medium on dimensionless variables it can be represented as

$$p + \Phi + \frac{U^2}{2} + \frac{\partial \varphi}{\partial t} = f(t), \qquad (5)$$

where U is the velocity module, $U = \sqrt{u^2 + v^2 + w^2}$;

φ is the velocity potential;

$f(t)$ is an arbitrary function of time.

The Lagrange — Cauchy integral is just for the potential motion both the ideal and viscous medium.

In the case of steady - state motion of an ideal medium along the stream line holds the Bernoulli's integral. In the case of an incompressible medium on dimensionless variables it can be written as

$$p + \Phi + \frac{U^2}{2} = C_{sl}, \qquad (6)$$

where C_{sl} denotes a constant depending on the choice of the stream line.

In some cases constant C_{sl} on the right-hand side does not depend on choice of stream line. This is the case if additional condition

$$\vec{U} \times \vec{\Omega} = 0, \qquad (7)$$

where $\vec{U}$ is the velocity vector, $\vec{\Omega}$ is the swirl vector, " $\times$ " is the sign of vector product.

In this case integral is often called as the Euler — Bernoulli integral. It can be set by equality as the next

$$p + \Phi + \frac{U^2}{2} = C, \tag{8}$$

where C denotes an absolute constant not depending on the choice of stream line.

Integrals (5), (6) and (8) be called one of the basic relations of classical fluid mechanics. A lot of diverse tasks were decided with their help. But even for these well-known integrals not all questions clarified through. For example it is not clear how the Bernoulli's integral (6) transforms, if the characteristic point is not to take along the stream line, but as a different way. The ratio of (6) will be broken and it is not clear what will happen instead. It is also not clear what will be instead of (5) if we abandon condition of potentiality of motion.

Each of mentioned integrals contains the same combination of $p + \Phi + \frac{U^2}{2}$ which is usually called as trinomial of Bernoulli. For this reason, sometimes all three of these integrals are called as Bernoulli integrals, although this is not entirely accurate, since the initial assumptions and the range of applicability of integrals (5), (6), (8) are different.

Classical integrals (5), (6), (8) were obtained more than two hundred years ago. Since then, numerous attempts have been made to construct new integrals [5-7]. But the proposed integrals covered only particular cases and related only to a specific cases of motion. The question arises about the existence of a general root integral just for all space of variables. So that this root integral also would contain the Bernoulli's trinomial and would cover the integrals (5), (6), (8), as a special cases. This situation can be illustrated by Figure 1.

In the diagram shown in Figure 1 integrals of Bernoulli, Euler — Bernoulli and Lagrange — Cauchy denoted as B, E and L in accordance and unknown root integral denoted with a question mark.

2. METHODS

The proposed research is based on methods of partial differential equations and mathematical physics. The first three equations of (1-4) are nonlinear but they are the same type since they can be reduced to free divergent form

Integrals of the Navier – Stokes and Euler Equations ...

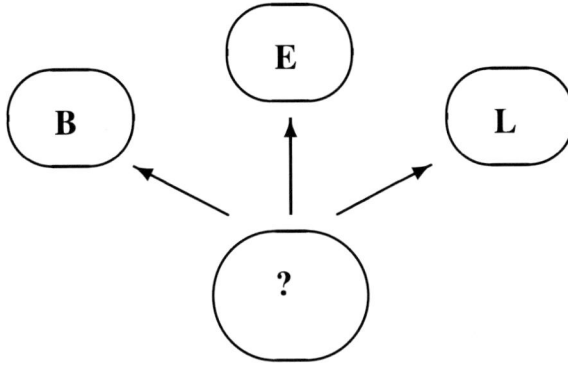

Figure 1. Known integrals of motion for an incompressible medium and unknown root integral.

as

$$\frac{\partial}{\partial x}(u^2 + p + \Phi - \frac{1}{Re}\frac{\partial u}{\partial x}) + \frac{\partial}{\partial y}(uv - \frac{1}{Re}\frac{\partial u}{\partial y}) + \frac{\partial}{\partial z}(uw - \frac{1}{Re}\frac{\partial u}{\partial z}) + \frac{\partial u}{\partial t} = 0,$$

$$\frac{\partial}{\partial x}(uv - \frac{1}{Re}\frac{\partial v}{\partial x}) + \frac{\partial}{\partial y}(v^2 + p + \Phi - \frac{1}{Re}\frac{\partial v}{\partial y}) + \frac{\partial}{\partial z}(vw - \frac{1}{Re}\frac{\partial v}{\partial z}) + \frac{\partial v}{\partial t} = 0,$$

$$\frac{\partial}{\partial x}(uw - \frac{1}{Re}\frac{\partial w}{\partial x}) + \frac{\partial}{\partial y}(vw - \frac{1}{Re}\frac{\partial w}{\partial y}) + \frac{\partial}{\partial z}(w^2 + p + \Phi - \frac{1}{Re}\frac{\partial w}{\partial z}) + \frac{\partial w}{\partial t} = 0.$$

Equation (4) has a free divergent form initially. Therefore, we can state that each of the equations of system (1-4) is represented in the form

$$\frac{\partial P_i}{\partial x} + \frac{\partial Q_i}{\partial y} + \frac{\partial R_i}{\partial z} + \frac{\partial S_i}{\partial t} = 0, \qquad (9)$$

where P_i, Q_i, R_i, S_i are some combinations of unknowns u, v, w, p and their first derivatives on spatial coordinates.

Using the methods of partial differential equations, it can be shown that each equation of the form (9) allows the integration. The resulting ratio can be combined and simplified by full or partial exclusion of non-divergent terms. Some of the resulting relationships can be brought to the divergence form (9) and then integrated again. Transformations leads to nine basic ratios. Using the simplest notation they can be represented as [8-11]

$$p - p_0 + \Phi + \frac{U^2}{2} + d + d_t = \alpha_4 + \beta_4 + \gamma_4, \tag{10}$$

$$u^2 - v^2 + \frac{2}{Re}\left(-\frac{\partial u}{\partial x} + \frac{\partial v}{\partial y}\right) = -\frac{\partial^2 \Psi_{10}}{\partial x^2} + \frac{\partial^2 \Psi_{10}}{\partial y^2} - \frac{\partial^2 \Psi_{11}}{\partial z^2} - \frac{\partial^2 \Psi_{12}}{\partial z^2} + \frac{\partial^2 \Psi_{15}}{\partial y \partial z} +$$

$$\frac{\partial^2 \Psi_{14}}{\partial x \partial z} + \frac{\partial}{\partial t}\left(-\frac{\partial \Psi_1}{\partial x} + \frac{\partial \Psi_3}{\partial y} + \frac{\partial (\Psi_5 + \Psi_6)}{\partial z}\right) + 3(\alpha_4 - \beta_4), \tag{11}$$

$$v^2 - w^2 + \frac{2}{Re}\left(-\frac{\partial v}{\partial y} + \frac{\partial w}{\partial z}\right) = \frac{\partial^2 \Psi_{10}}{\partial x^2} + \frac{\partial^2 \Psi_{11}}{\partial x^2} - \frac{\partial^2 \Psi_{12}}{\partial y^2} + \frac{\partial^2 \Psi_{12}}{\partial z^2} - \frac{\partial^2 \Psi_{13}}{\partial x \partial y} -$$

$$\frac{\partial^2 \Psi_{14}}{\partial x \partial z} + \frac{\partial}{\partial t}\left(\frac{\partial (\Psi_1 + \Psi_2)}{\partial x} + \frac{\partial \Psi_4}{\partial y} - \frac{\partial \Psi_6}{\partial z}\right) + 3(\beta_4 - \gamma_4), \tag{12}$$

$$uv - \frac{1}{Re}\left(\frac{\partial v}{\partial x} + \frac{\partial u}{\partial y}\right) = -\frac{\partial^2 \Psi_{10}}{\partial x \partial y} + \frac{1}{2}\frac{\partial}{\partial z}\left(-\frac{\partial \Psi_{15}}{\partial x} + \frac{\partial \Psi_{14}}{\partial y} + \frac{\partial \Psi_{13}}{\partial z}\right) +$$

$$\frac{1}{2}\frac{\partial}{\partial t}\left(-\frac{\partial \Psi_3}{\partial x} - \frac{\partial \Psi_1}{\partial y} - \frac{\partial (\Psi_8 + \Psi_9)}{\partial z}\right) + \frac{1}{2}\left(\frac{\partial \alpha_1}{\partial z} - \frac{\partial \alpha_3}{\partial t} + \frac{\partial \beta_1}{\partial z} - \frac{\partial \beta_2}{\partial t}\right), \tag{13}$$

$$uw - \frac{1}{Re}\left(\frac{\partial w}{\partial x} + \frac{\partial u}{\partial z}\right) = \frac{\partial^2 \Psi_{11}}{\partial x \partial z} + \frac{1}{2}\frac{\partial}{\partial y}\left(-\frac{\partial \Psi_{15}}{\partial x} - \frac{\partial \Psi_{14}}{\partial y} - \frac{\partial \Psi_{13}}{\partial z}\right) +$$

$$\frac{1}{2}\frac{\partial}{\partial t}\left(-\frac{\partial \Psi_5}{\partial x} + \frac{\partial (\Psi_9 - \Psi_7)}{\partial y} + \frac{\partial \Psi_2}{\partial z}\right) - \frac{1}{2}\left(\frac{\partial \alpha_1}{\partial y} + \frac{\partial \alpha_2}{\partial t} - \frac{\partial \gamma_1}{\partial y} + \frac{\partial \gamma_3}{\partial t}\right), \tag{14}$$

$$vw - \frac{1}{Re}\left(\frac{\partial w}{\partial y} + \frac{\partial v}{\partial z}\right) = -\frac{\partial^2 \Psi_{12}}{\partial y \partial z} + \frac{1}{2}\frac{\partial}{\partial x}\left(\frac{\partial \Psi_{14}}{\partial y} + \frac{\partial \Psi_{15}}{\partial x} - \frac{\partial \Psi_{13}}{\partial z}\right) +$$

$$\frac{1}{2}\frac{\partial}{\partial t}\left(\frac{\partial (\Psi_7 + \Psi_8)}{\partial x} + \frac{\partial \Psi_6}{\partial y} + \frac{\partial \Psi_4}{\partial z}\right) - \frac{1}{2}\left(\frac{\partial \beta_1}{\partial x} + \frac{\partial \beta_3}{\partial t} + \frac{\partial \gamma_1}{\partial x} + \frac{\partial \gamma_2}{\partial t}\right), \tag{15}$$

$$u = \frac{1}{2}\left(\frac{\partial}{\partial y}\left(-\frac{\partial \Psi_3}{\partial x} + \frac{\partial \Psi_1}{\partial y} + \frac{\partial \Psi_7}{\partial z}\right) + \frac{\partial}{\partial z}\left(-\frac{\partial \Psi_5}{\partial x} + \frac{\partial \Psi_8}{\partial y} - \frac{\partial \Psi_2}{\partial z}\right)\right) +$$

$$\frac{1}{2}\left(\frac{\partial \alpha_2}{\partial z} + \frac{\partial \alpha_3}{\partial y} + \frac{\partial \delta_1}{\partial y} + \frac{\partial \delta_2}{\partial z}\right), \tag{16}$$

$$v = \frac{1}{2}\frac{\partial}{\partial x}(\frac{\partial \Psi_3}{\partial x} - \frac{\partial \Psi_1}{\partial y} - \frac{\partial \Psi_7}{\partial z}) + \frac{\partial}{\partial z}(\frac{\partial \Psi_9}{\partial x} + \frac{\partial \Psi_6}{\partial y} - \frac{\partial \Psi_4}{\partial z})) +$$
$$\frac{1}{2}(\frac{\partial \beta_2}{\partial x} + \frac{\partial \beta_3}{\partial z} - \frac{\partial \delta_1}{\partial x} + \frac{\partial \delta_3}{\partial z}), \qquad (17)$$

$$w = \frac{1}{2}\frac{\partial}{\partial x}(\frac{\partial \Psi_5}{\partial x} - \frac{\partial \Psi_8}{\partial y} + \frac{\partial \Psi_2}{\partial z}) + \frac{\partial}{\partial y}(-\frac{\partial \Psi_9}{\partial x} - \frac{\partial \Psi_6}{\partial y} + \frac{\partial \Psi_4}{\partial z})) +$$
$$\frac{1}{2}(\frac{\partial \gamma_2}{\partial y} + \frac{\partial \gamma_3}{\partial x} - \frac{\partial \delta_2}{\partial x} - \frac{\partial \delta_3}{\partial y}). \qquad (18)$$

Note that the new values Ψ_i appears on ratios of (10-18). These functions are the associated unknown arising as a result of integration. These new unknowns are complemented the system of unknowns. Name of the "stream pseudo functions" introduced for them.

Values α_k, β_k, γ_k, δ_k denotes an arbitrary function on three variables only. These functions and their first derivatives presents in the right-hand side of equation (10-18) additively. Each function α_k, β_k, γ_k, δ_k does not depend on one of variables respectively x, y, z or t, so as the following equalities must hold

$$\frac{\partial \alpha_k}{\partial x} = \frac{\partial \beta_k}{\partial y} = \frac{\partial \gamma_k}{\partial z} = \frac{\partial \delta_k}{\partial t} = 0. \qquad (19)$$

Value of p_0, U, d and d_t presents on (10). Meaning of the first two quantities is obvious: p_0 is an arbitrary pressure constant and U is the velocity modulus. Value d and d_t corresponds to the reduced dissipation, respectively stationary and non-stationary and defined by formulas

$$d = -\frac{U^2}{6} - \frac{1}{3}(\Delta_{xy}\Psi_{10} - \Delta_{xz}\Psi_{11} + \Delta_{yz}\Psi_{12} + \frac{\partial^2 \Psi_{13}}{\partial x \partial y} - \frac{\partial^2 \Psi_{14}}{\partial x \partial z} + \frac{\partial^2 \Psi_{15}}{\partial y \partial z}), \qquad (20)$$

$$d_t = \frac{1}{3}\frac{\partial}{\partial t}(\frac{\partial(\Psi_2 - \Psi_1)}{\partial x} + \frac{\partial(\Psi_4 - \Psi_3)}{\partial y} + \frac{\partial(\Psi_6 - \Psi_5)}{\partial z}). \qquad (21)$$

Characters of Δ_{yz}, Δ_{xz}, Δ_{xy} on (21) denotes the partial Laplace operators on spatial coordinates

$$\Delta_{yz} = \frac{\partial^2}{\partial y^2} + \frac{\partial^2}{\partial z^2}, \quad \Delta_{xz} = \frac{\partial^2}{\partial x^2} + \frac{\partial^2}{\partial z^2}, \quad \Delta_{xy} = \frac{\partial^2}{\partial x^2} + \frac{\partial^2}{\partial y^2}.$$

The relations (10-18) connect the main unknowns u, v, w, p, associated ones Ψ_i, and an arbitrary functions of three variables α_k, β_k, γ_k, δ_k. The order of derivatives for major unknowns is one less than their order in the original equations (1-4). So the relations (10-18) considered together constitute the first integral of Navier — Stokes equations (1-4). This statement is just in general case without depending on configuration of domain of variables x, y, z, t, value of Reynolds number Re, boundary and initial conditions and other factors of private character. The conclusion and the proof of this statement is presented in [10].

The fact that integral is represented by nine interdependent relationships rather than one should not be surprised. When lifting the restrictions of a private nature, such as the potentiality of motion, an ideality of medium, the motion along the stream line, then all this should lead to a complication of resulting ratios. It's obvious that number of resulting ratios increases and their structure is complicates.

The emergence of new associated unknowns Ψ_i in ratios (10-18) should not surprise as well. Analogues of associated unknowns present in known integrals of (5) and (6). But they present under different name and under different designations. On integral (5) the role of an associated unknown plays the potential of velocity φ. On integral of (6) the associated unknown is the value of the right-hand side C_{sl} since this value depends on the stream line and therefore is a function. On integral (5) an arbitrary additive function presents in the form of $f(t)$.

Note that similar reasoning and study methodology were used in [12] as well. In this paper the equation containing an associated unknown called as "parametrically defined" differential equations.

3. Results

So, the first integral of $3D$ Navier — Stokes equations (1-4) just for all space of variables is represented by nine interconnected relations (10-18). This integral represents the desired root integral, which was mentioned in the introduction. Comparison of this integral with known of Lagrange — Cauchy (5), Bernoulli (6) and Euler — Bernoulli (8) integrals leads to the statement as follows. Each of integrals (5), (6), (8) is a special case of (10-18). We consider each of these statements in more detail for motion of incompressible medium.

3.1. Lagrange — Cauchy Integral as the Special Case of the Root Integral

The following theorem is valid.

Theorem 3.1. *Integral of Lagrange — Cauchy (5) is a special case of the root integral presented by ratios (10-18).*

Proof of the Theorem 3.1. Consider ratios (10-18) and suppose that medium flow is of potential. Then the potential of velocity φ is exist and it satisfies the Laplace equation $\Delta\varphi = 0$. The following equality hold [1-2]

$$u = \frac{\partial\varphi}{\partial x}, \quad v = \frac{\partial\varphi}{\partial y}, \quad w = \frac{\partial\varphi}{\partial z}. \tag{22}$$

Each of velocity components satisfies the Laplace equation as well

$$\Delta u = 0, \quad \Delta v = 0, \quad \Delta w = 0. \tag{23}$$

Motion is irrotational so $\vec{\Omega} = 0$. The latter is equivalent to three scalar equations as the next

$$\frac{\partial w}{\partial y} - \frac{\partial v}{\partial z} = 0, \quad \frac{\partial u}{\partial z} - \frac{\partial w}{\partial x} = 0, \quad \frac{\partial v}{\partial x} - \frac{\partial u}{\partial y} = 0. \tag{24}$$

For the case under consideration all ratio of (22-24) done. We have to prove that under these conditions from (10-18) should be (5).

Ratio (10) allocated from the (10-18). It connects the largest number of unknown and only in it appears the pressure p. The left-hand side of (10) is differs from Bernoulli's trinomial on the value of $d + d_t$. Let's calculate first derivatives of $d + d_t$ with respect to spatial coordinates.

3.1.1. Considering the derivative with respect to ∂x due to (20), (21) we have the equality

$$\frac{\partial(d+d_t)}{\partial x} = -\frac{1}{3}(u\frac{\partial u}{\partial x}+v\frac{\partial v}{\partial x}+w\frac{\partial w}{\partial x}) + \frac{1}{3}(-\frac{\partial^3\Psi_{10}}{\partial x^3} - \frac{\partial^3\Psi_{10}}{\partial x \partial y^2} + \frac{\partial^3\Psi_{11}}{\partial x^3} +$$

$$\frac{\partial^3 \Psi_{11}}{\partial x \partial z^2} - \frac{\partial^3 \Psi_{12}}{\partial x \partial y^2} - \frac{\partial^3 \Psi_{12}}{\partial x \partial z^2} - \frac{\partial^3 \Psi_{13}}{\partial x^2 \partial y} + \frac{\partial^3 \Psi_{14}}{\partial x^2 \partial z} - \frac{\partial^3 \Psi_{15}}{\partial x \partial y \partial z}) + \frac{1}{3}\frac{\partial}{\partial t}(\frac{\partial^2 \Psi_2}{\partial x^2} -$$

$$\frac{\partial^2 \Psi_1}{\partial x^2} + \frac{\partial^2 \Psi_4}{\partial x \partial y} - \frac{\partial^2 \Psi_3}{\partial x \partial y} + \frac{\partial^2 \Psi_6}{\partial x \partial z} - \frac{\partial^2 \Psi_5}{\partial x \partial z}). \tag{25}$$

The second group of members of right-hand side is the most difficult one. It contains the third derivatives with respect to spatial coordinates. Let's transform this fragment using some of ratio (11-18). We calculate $\frac{2}{3}\frac{\partial}{\partial x}$ from (11), $\frac{1}{3}\frac{\partial}{\partial x}$ from (12), $\frac{\partial}{\partial y}$ from (13), $\frac{\partial}{\partial z}$ from (14). Combining the results, we obtain an expression for desired fragment with third derivatives in the form of

$$\frac{1}{3}(-\frac{\partial^3 \Psi_{10}}{\partial x^3} - \frac{\partial^3 \Psi_{10}}{\partial x \partial y^2} + \frac{\partial^3 \Psi_{11}}{\partial x^3} + \frac{\partial^3 \Psi_{11}}{\partial x \partial z^2} - \frac{\partial^3 \Psi_{12}}{\partial x \partial y^2} - \frac{\partial^3 \Psi_{12}}{\partial x \partial z^2} - \frac{\partial^3 \Psi_{13}}{\partial x^2 \partial y} + \frac{\partial^3 \Psi_{14}}{\partial x^2 \partial z} -$$

$$\frac{\partial^3 \Psi_{15}}{\partial x \partial y \partial z}) = \frac{4}{3}u\frac{\partial u}{\partial x} - \frac{2}{3}v\frac{\partial v}{\partial x} - \frac{2}{3}w\frac{\partial w}{\partial x} + v\frac{\partial u}{\partial y} + u\frac{\partial v}{\partial y} + w\frac{\partial u}{\partial z} + u\frac{\partial w}{\partial z} +$$

$$\frac{1}{Re}(-\frac{4}{3}\frac{\partial^2 u}{\partial x^2} - \frac{1}{3}\frac{\partial^2 v}{\partial x \partial y} - \frac{1}{3}\frac{\partial^2 w}{\partial x \partial z} - \frac{\partial^2 u}{\partial y^2} - \frac{\partial^2 u}{\partial z^2}) + \frac{\partial}{\partial t}(\frac{1}{3}\frac{\partial^2 \Psi_1}{\partial x^2} + \frac{1}{2}\frac{\partial^2 \Psi_1}{\partial y^2} -$$

$$\frac{1}{3}\frac{\partial^2 \Psi_2}{\partial x^2} - \frac{1}{2}\frac{\partial^2 \Psi_2}{\partial z^2} - \frac{1}{6}\frac{\partial^2 \Psi_3}{\partial x \partial y} - \frac{1}{3}\frac{\partial^2 \Psi_4}{\partial x \partial y} - \frac{1}{6}\frac{\partial^2 \Psi_5}{\partial x \partial z} -$$

$$\frac{1}{3}\frac{\partial^2 \Psi_6}{\partial x \partial z} + \frac{1}{2}\frac{\partial^2 \Psi_7}{\partial y \partial z} + \frac{1}{2}\frac{\partial^2 \Psi_8}{\partial y \partial z}) + \frac{1}{2}\frac{\partial^2 \alpha_2}{\partial z \partial t} + \frac{1}{2}\frac{\partial^2 \alpha_3}{\partial y \partial t} + \frac{\partial \beta_4}{\partial x} + \frac{\partial \gamma_4}{\partial x}. \tag{26}$$

Transform (25) according to (26) we obtain equality as the next

$$\frac{\partial(d + d_t)}{\partial x} = \frac{1}{Re}(-\Delta u - \frac{1}{3}\frac{\partial}{\partial x}(\frac{\partial u}{\partial x} + \frac{\partial v}{\partial y} + \frac{\partial w}{\partial z})) + u(\frac{\partial u}{\partial x} + \frac{\partial v}{\partial y} + \frac{\partial w}{\partial z}) +$$

$$v(\frac{\partial u}{\partial y} - \frac{\partial v}{\partial x}) + w(\frac{\partial u}{\partial z} - \frac{\partial w}{\partial x}) + \frac{1}{2}\frac{\partial}{\partial t}(\frac{\partial^2 \Psi_1}{\partial y^2} - \frac{\partial^2 \Psi_2}{\partial z^2} - \frac{\partial^2 \Psi_3}{\partial x \partial y} - \frac{\partial^2 \Psi_5}{\partial x \partial z} + \tag{27}$$

$$\frac{\partial^2 \Psi_7}{\partial y \partial z} + \frac{\partial^2 \Psi_8}{\partial y \partial z}) + \frac{1}{2}\frac{\partial^2 \alpha_2}{\partial z \partial t} + \frac{1}{2}\frac{\partial^2 \alpha_3}{\partial y \partial t} + \frac{\partial \beta_4}{\partial x} + \frac{\partial \gamma_4}{\partial x}.$$

The last relation allows significant simplification. From (23) it follows that $\Delta u = 0$. Taking into account the continuity equation (4), we can conclude that group of terms proportional to $\frac{1}{Re}$ vanishes.

Integrals of the Navier – Stokes and Euler Equations ... 71

The second group on right-hand side of (27) vanishes as well in consequence of (4) and (24). We use now the equation (16) and one of (19) in the form of $\frac{\partial \delta_k}{\partial t} = 0$. As a result we state that all the remaining terms on the right-hand side of (27) are reduced to $\frac{\partial u}{\partial t} + \frac{\partial \beta_4}{\partial x} + \frac{\partial \gamma_4}{\partial x}$.

Then equality (27) takes the form of

$$\frac{\partial (d + d_t)}{\partial x} = \frac{\partial u}{\partial t} + \frac{\partial \beta_4}{\partial x} + \frac{\partial \gamma_4}{\partial x}. \tag{28}$$

3.1.2. We find the first derivative with respect to ∂y value of $d + d_t$. In view of (20), (21) we obtain the original equation in the form of

$$\frac{\partial (d + d_t)}{\partial y} = -\frac{1}{3}(u\frac{\partial u}{\partial y} + v\frac{\partial v}{\partial y} + w\frac{\partial w}{\partial y}) + \frac{1}{3}(-\frac{\partial^3 \Psi_{10}}{\partial x^2 \partial y} - \frac{\partial^3 \Psi_{10}}{\partial y^3} + \frac{\partial^3 \Psi_{11}}{\partial x^2 \partial y} +$$

$$\frac{\partial^3 \Psi_{11}}{\partial y \partial z^2} - \frac{\partial^3 \Psi_{12}}{\partial y^3} - \frac{\partial^3 \Psi_{12}}{\partial y \partial z^2} + \frac{\partial^3 \Psi_{13}}{\partial x \partial y^2} + \frac{\partial^3 \Psi_{14}}{\partial x \partial y \partial z} - \frac{\partial^3 \Psi_{15}}{\partial y^2 \partial z}) + \frac{1}{3}\frac{\partial}{\partial t}(\frac{\partial^2 \Psi_2}{\partial x \partial y} -$$

$$\frac{\partial^2 \Psi_1}{\partial x \partial y} + \frac{\partial^2 \Psi_4}{\partial y^2} - \frac{\partial^2 \Psi_3}{\partial y^2} + \frac{\partial^2 \Psi_6}{\partial y \partial z} - \frac{\partial^2 \Psi_5}{\partial y \partial z}). \tag{29}$$

Let's simplify the fragment with third derivatives of right-hand side. We find in series $-\frac{1}{3}\frac{\partial}{\partial y}$ of (11), $\frac{1}{3}\frac{\partial}{\partial y}$ of (12), $\frac{\partial}{\partial x}$ of (13), $\frac{\partial}{\partial z}$ of (15). Folding the results we get an expression

$$\frac{1}{3}(-\frac{\partial^3 \Psi_{10}}{\partial x^2 \partial y} - \frac{\partial^3 \Psi_{10}}{\partial y^3} + \frac{\partial^3 \Psi_{11}}{\partial x^2 \partial y} + \frac{\partial^3 \Psi_{11}}{\partial y \partial z^2} - \frac{\partial^3 \Psi_{12}}{\partial y^3} - \frac{\partial^3 \Psi_{12}}{\partial y \partial z^2} + \frac{\partial^3 \Psi_{13}}{\partial x \partial y^2} +$$

$$\frac{\partial^3 \Psi_{14}}{\partial x \partial y \partial z} - \frac{\partial^3 \Psi_{15}}{\partial y^2 \partial z}) = \frac{4}{3}v\frac{\partial v}{\partial y} - \frac{2}{3}u\frac{\partial u}{\partial y} - \frac{2}{3}w\frac{\partial w}{\partial y} + v\frac{\partial u}{\partial x} + u\frac{\partial v}{\partial x} + w\frac{\partial v}{\partial z} + v\frac{\partial w}{\partial z} +$$

$$\frac{1}{Re}(-\frac{4}{3}\frac{\partial^2 v}{\partial y^2} - \frac{1}{3}\frac{\partial^2 u}{\partial x \partial y} - \frac{1}{3}\frac{\partial^2 w}{\partial y \partial z} - \frac{\partial^2 v}{\partial x^2} - \frac{\partial^2 v}{\partial z^2}) + \frac{\partial}{\partial t}(-\frac{1}{6}\frac{\partial^2 \Psi_1}{\partial x \partial y} - \frac{1}{3}\frac{\partial^2 \Psi_2}{\partial x \partial y} +$$

$$\frac{1}{2}\frac{\partial^2 \Psi_3}{\partial x^2} + \frac{1}{3}\frac{\partial^2 \Psi_3}{\partial y^2} - \frac{1}{3}\frac{\partial^2 \Psi_4}{\partial y^2} - \frac{1}{2}\frac{\partial^2 \Psi_4}{\partial z^2} + \frac{1}{3}\frac{\partial^2 \Psi_5}{\partial y \partial z} +$$

$$\frac{1}{6}\frac{\partial^2 \Psi_6}{\partial y \partial z} - \frac{1}{2}\frac{\partial^2 \Psi_7}{\partial x \partial z} + \frac{1}{2}\frac{\partial^2 \Psi_9}{\partial x \partial z}) + \frac{1}{2}\frac{\partial^2 \beta_2}{\partial x \partial t} + \frac{1}{2}\frac{\partial^2 \beta_3}{\partial z \partial t} + \frac{\partial \beta_4}{\partial y} + \frac{\partial \gamma_4}{\partial y}. \tag{30}$$

Convert of (29) due to (30) leads to equality as the next

$$\frac{\partial (d + d_t)}{\partial y} = \frac{1}{Re}(-\Delta v - \frac{1}{3}\frac{\partial}{\partial y}(\frac{\partial u}{\partial x} + \frac{\partial v}{\partial y} + \frac{\partial w}{\partial z})) + v(\frac{\partial u}{\partial x} + \frac{\partial v}{\partial y} + \frac{\partial w}{\partial z}) +$$

$$u(\frac{\partial v}{\partial x} - \frac{\partial u}{\partial y}) + w(\frac{\partial v}{\partial z} - \frac{\partial w}{\partial y}) + \frac{1}{2}\frac{\partial}{\partial t}(-\frac{\partial^2 \Psi_1}{\partial x \partial y} + \frac{\partial^2 \Psi_3}{\partial x^2} - \frac{\partial^2 \Psi_4}{\partial z^2} + \qquad (31)$$

$$\frac{\partial^2 \Psi_6}{\partial y \partial z} - \frac{\partial^2 \Psi_7}{\partial x \partial z} + \frac{\partial^2 \Psi_9}{\partial x \partial z}) + \frac{1}{2}\frac{\partial^2 \beta_2}{\partial x \partial t} + \frac{1}{2}\frac{\partial^2 \beta_3}{\partial z \partial t} + \frac{\partial \alpha_4}{\partial y} + \frac{\partial \gamma_4}{\partial y}.$$

Let us simplify (31) based on the initial assumptions. Using (23) together with the continuity equation (4) we state that first group of terms on the right-hand side of (31) vanishes.

Note that as a consequence of equation (4) and conditions of (24) the second group of terms is vanishes as well. If we use (17) and (19) then all the remaining terms on the right-hand side are reduced to $\frac{\partial v}{\partial t} + \frac{\partial \alpha_4}{\partial y} + \frac{\partial \gamma_4}{\partial y}$.

As a result equality (31) takes the form

$$\frac{\partial (d + d_t)}{\partial y} = \frac{\partial v}{\partial t} + \frac{\partial \alpha_4}{\partial y} + \frac{\partial \gamma_4}{\partial y}. \qquad (32)$$

3.1.3. Let's find the first derivative on ∂z value of $d + d_t$. Given (20), (21) as the source we have the equality as the next

$$\frac{\partial(d+d_t)}{\partial z} = -\frac{1}{3}(u\frac{\partial u}{\partial z} + v\frac{\partial v}{\partial z} + w\frac{\partial w}{\partial z}) + \frac{1}{3}(-\frac{\partial^3 \Psi_{10}}{\partial x^2 \partial z} - \frac{\partial^3 \Psi_{10}}{\partial y^2 \partial z} + \frac{\partial^3 \Psi_{11}}{\partial x^2 \partial z} +$$

$$\frac{\partial^3 \Psi_{11}}{\partial z^3} - \frac{\partial^3 \Psi_{12}}{\partial y^2 \partial z} - \frac{\partial^3 \Psi_{12}}{\partial z^3} + \frac{\partial^3 \Psi_{13}}{\partial x \partial y \partial z} + \frac{\partial^3 \Psi_{14}}{\partial x \partial z^2} - \frac{\partial^3 \Psi_{15}}{\partial y \partial z^2}) + \frac{1}{3}\frac{\partial}{\partial t}(\frac{\partial^2 \Psi_2}{\partial x \partial z} -$$

$$\frac{\partial^2 \Psi_1}{\partial x \partial z} + \frac{\partial^2 \Psi_4}{\partial y \partial z} - \frac{\partial^2 \Psi_3}{\partial y \partial z} + \frac{\partial^2 \Psi_6}{\partial z^2} - \frac{\partial^2 \Psi_5}{\partial z^2}). \qquad (33)$$

A group of members with the third derivatives on the right-hand side of (33) can be converted using (11-15). Let's calculate $-\frac{1}{3}\frac{\partial}{\partial z}$ of (11), $-\frac{2}{3}\frac{\partial}{\partial z}$ of (12), $\frac{\partial}{\partial x}$ of (14), $\frac{\partial}{\partial y}$ of (15) and the results add up. We obtain an expression for needed fragment with the third derivatives in the form of

$$\frac{1}{3}(-\frac{\partial^3 \Psi_{10}}{\partial x^2 \partial z} - \frac{\partial^3 \Psi_{10}}{\partial y^2 \partial z} + \frac{\partial^3 \Psi_{11}}{\partial x^2 \partial z} + \frac{\partial^3 \Psi_{11}}{\partial z^3} - \frac{\partial^3 \Psi_{12}}{\partial z^3} - \frac{\partial^3 \Psi_{12}}{\partial y \partial z^2} - \frac{\partial^3 \Psi_{13}}{\partial x \partial y \partial z} +$$

$$\frac{\partial^3 \Psi_{14}}{\partial x \partial z^2} - \frac{\partial^3 \Psi_{15}}{\partial y \partial z^2}) = -\frac{2}{3}u\frac{\partial u}{\partial z} - \frac{2}{3}v\frac{\partial v}{\partial z} + \frac{4}{3}w\frac{\partial w}{\partial z} + w\frac{\partial u}{\partial x} + u\frac{\partial w}{\partial x} + w\frac{\partial v}{\partial y} + v\frac{\partial w}{\partial y} +$$

$$\frac{1}{Re}(-\frac{1}{3}\frac{\partial^2 v}{\partial y \partial z} - \frac{1}{3}\frac{\partial^2 u}{\partial x \partial z} - \frac{4}{3}\frac{\partial^2 w}{\partial z^2} - \frac{\partial^2 w}{\partial x^2} - \frac{\partial^2 w}{\partial y^2}) + \frac{\partial}{\partial t}(\frac{1}{3}\frac{\partial^2 \Psi_1}{\partial x \partial z} + \frac{1}{6}\frac{\partial^2 \Psi_2}{\partial x \partial z} +$$

$$\frac{1}{3}\frac{\partial^2 \Psi_3}{\partial y \partial z} + \frac{1}{6}\frac{\partial^2 \Psi_4}{\partial y \partial z} + \frac{1}{3}\frac{\partial^2 \Psi_5}{\partial z^2} + \frac{1}{2}\frac{\partial^2 \Psi_5}{\partial x^2} - \frac{1}{3}\frac{\partial^2 \Psi_6}{\partial z^2} -$$

$$\frac{1}{2}\frac{\partial^2 \Psi_6}{\partial y^2} - \frac{1}{2}\frac{\partial^2 \Psi_8}{\partial x \partial y} - \frac{1}{2}\frac{\partial^2 \Psi_9}{\partial x \partial y}) + \frac{1}{2}\frac{\partial^2 \gamma_2}{\partial y \partial t} + \frac{1}{2}\frac{\partial^2 \gamma_3}{\partial x \partial t} + \frac{\partial \alpha_4}{\partial z} + \frac{\partial \beta_4}{\partial z}. \qquad (34)$$

Converting (33) due to (34) we arrive to the next

$$\frac{\partial (d + d_t)}{\partial z} = \frac{1}{Re}(-\Delta w - \frac{1}{3}\frac{\partial}{\partial z}(\frac{\partial u}{\partial x} + \frac{\partial v}{\partial y} + \frac{\partial w}{\partial z})) + w(\frac{\partial u}{\partial x} + \frac{\partial v}{\partial y} + \frac{\partial w}{\partial z}) +$$

$$u(\frac{\partial w}{\partial x} - \frac{\partial u}{\partial z}) + v(\frac{\partial w}{\partial y} - \frac{\partial v}{\partial z}) + \frac{1}{2}\frac{\partial}{\partial t}(\frac{\partial^2 \Psi_2}{\partial x \partial z} + \frac{\partial^2 \Psi_4}{\partial y \partial z} + \frac{\partial^2 \Psi_5}{\partial x^2} - \qquad (35)$$

$$\frac{\partial^2 \Psi_6}{\partial y^2} - \frac{\partial^2 \Psi_8}{\partial x \partial y} - \frac{\partial^2 \Psi_9}{\partial x \partial y}) + \frac{1}{2}\frac{\partial^2 \gamma_2}{\partial y \partial t} + \frac{1}{2}\frac{\partial^2 \gamma_3}{\partial x \partial t} + \frac{\partial \alpha_4}{\partial z} + \frac{\partial \beta_4}{\partial z}.$$

We use the continuity equation (4) together with (23) and state that the terms proportional to $\frac{1}{Re}$ vanishes. The second group of terms is also vanishes due to equation (4) and conditions (24). Using the equations (18) and (19) we note that remaining terms of the right-hand side of (35) are reduced to $\frac{\partial w}{\partial t} + \frac{\partial \alpha_4}{\partial z} + \frac{\partial \beta_4}{\partial z}$.

As a result, equality of (35) takes the form

$$\frac{\partial (d + d_t)}{\partial z} = \frac{\partial w}{\partial t} + \frac{\partial \alpha_4}{\partial z} + \frac{\partial \beta_4}{\partial z}. \qquad (36)$$

3.1.4. So, the first derivatives with respect to coordinates x, y, z value of $d + d_t$ are defined by (28), (32), (36). These equations can be written in a more convenient form if we add in the right-hand sides for one term, respectively $\frac{\partial \alpha_4}{\partial x}$, $\frac{\partial \beta_4}{\partial y}$, $\frac{\partial \gamma_4}{\partial z}$. Each of these quantities is equal zero under the condition of (19), so the validity of (28), (32), (36) can not be broken. We use now the submission of u, v, w through potential φ according to (22) and change the order of differentiation with respect to time and coordinates in the second mixed derivatives. As a result of equality (28), (32), (36) takes the form

$$\frac{\partial (d + d_t)}{\partial x} = \frac{\partial}{\partial x}(\frac{\partial \varphi}{\partial t} + \alpha_4 + \beta_4 + \gamma_4),$$

$$\frac{\partial (d + d_t)}{\partial y} = \frac{\partial}{\partial y}(\frac{\partial \varphi}{\partial t} + \alpha_4 + \beta_4 + \gamma_4), \qquad (37)$$

$$\frac{\partial(d+d_t)}{\partial z} = \frac{\partial}{\partial z}\left(\frac{\partial \varphi}{\partial t} + \alpha_4 + \beta_4 + \gamma_4\right).$$

We state that value of $(d+d_t)$ and $\left(\frac{\partial \varphi}{\partial t}+\alpha_4+\beta_4+\gamma_4\right)$ have the same derivatives with respect to coordinates. Hence, they may differ only by an additive function of time, which can be chosen arbitrarily. We suppose that factor as $-f(t)$ and arrive to equality as the next

$$d + d_t = \frac{\partial \varphi}{\partial t} + \alpha_4 + \beta_4 + \gamma_4 - f(t), \tag{38}$$

Substituting (38) in (10) we obtain the equality equivalent to (5)

$$p + \Phi + \frac{U^2}{2} + \frac{\partial \varphi}{\partial t} - f(t) = 0.$$

Thus, the Lagrange — Cauchy integral (5) is a special case of root integral (10-18).

Note, that this result is valid regardless of value of the Reynolds number Re. Hence, the theorem is valid as for the case of motion of a viscous medium ($Re > 0$) and for the case of motion of an ideal medium ($Re = +\infty$).

The proof of Theorem **3.1** *is complete.*

3.2. Integral of Bernoulli as Special Case of the Root Integral

The following theorem are valid.

Theorem 3.2. Integral of Bernoulli (6) is a special case of the root integral (10-18).

Proof of the Theorem 3.2. We take as the assumptions those ones that form the basis for the derivation of Bernoulli's integral [1-2]. There are three of them. The medium is assumed an ideal, motion is steady - state and characteristic points are selected along the stream line. We prove that under these assumptions the Bernoulli's integral (6) is a consequence of (10-18).

According to the first two assumptions, we can assume that $\frac{1}{Re} = 0$ and $\frac{\partial}{\partial t} = 0$. The latter implies that $d_t = 0$ and ratio of (10) takes the form as

$$p - p_0 + \Phi + \frac{U^2}{2} + d = \alpha_4 + \beta_4 + \gamma_4. \tag{39}$$

The quantity p_0 is an additive constant and it can be taken into account on the right-hand side, since on the right-hand side we have an arbitrarily chosen functions. Thus the left-hand side of (39) differs from Bernoulli's trinomial on the value of d only. Let's find the derivatives of d with respect to spatial coordinates. To do this, you can use the ratio of (27), (31), (35). Note that these equations are simplified under given assumptions and take the form

$$\frac{\partial d}{\partial x} = u\left(\frac{\partial u}{\partial x} + \frac{\partial v}{\partial y} + \frac{\partial w}{\partial z}\right) + v\left(\frac{\partial u}{\partial y} - \frac{\partial v}{\partial x}\right) + w\left(\frac{\partial u}{\partial z} - \frac{\partial w}{\partial x}\right) + \frac{\partial \beta_4}{\partial x} + \frac{\partial \gamma_4}{\partial x},$$

$$\frac{\partial d}{\partial y} = v\left(\frac{\partial u}{\partial x} + \frac{\partial v}{\partial y} + \frac{\partial w}{\partial z}\right) + u\left(\frac{\partial v}{\partial x} - \frac{\partial u}{\partial y}\right) + w\left(\frac{\partial v}{\partial z} - \frac{\partial w}{\partial y}\right) + \frac{\partial \alpha_4}{\partial y} + \frac{\partial \gamma_4}{\partial y}, \tag{40}$$

$$\frac{\partial d}{\partial z} = w\left(\frac{\partial u}{\partial x} + \frac{\partial v}{\partial y} + \frac{\partial w}{\partial z}\right) + u\left(\frac{\partial w}{\partial x} - \frac{\partial u}{\partial z}\right) + v\left(\frac{\partial w}{\partial y} - \frac{\partial v}{\partial z}\right) + \frac{\partial \alpha_4}{\partial z} + \frac{\partial \beta_4}{\partial z}.$$

First term on the right-hand side vanishes due to continuity equation (4). To further simplify perform the following steps. Just as it was done in section **3.1.4**, add on right-hand side of (40) one term, respectively $\frac{\partial \alpha_4}{\partial x}$, $\frac{\partial \beta_4}{\partial y}$, $\frac{\partial \gamma_4}{\partial z}$. The validity of equality is not violated as each of these quantities equal zero. We reserve the nonlinear terms in the right-hand sides, and all others transfer to another one. Then (40) takes the form of

$$\frac{\partial (d - \alpha_4 - \beta_4 - \gamma_4)}{\partial x} = v\left(\frac{\partial u}{\partial y} - \frac{\partial v}{\partial x}\right) + w\left(\frac{\partial u}{\partial z} - \frac{\partial w}{\partial x}\right),$$

$$\frac{\partial (d - \alpha_4 - \beta_4 - \gamma_4)}{\partial y} = u\left(\frac{\partial v}{\partial x} - \frac{\partial u}{\partial y}\right) + w\left(\frac{\partial v}{\partial z} - \frac{\partial w}{\partial y}\right), \tag{41}$$

$$\frac{\partial (d - \alpha_4 - \beta_4 - \gamma_4)}{\partial z} = u\left(\frac{\partial w}{\partial x} - \frac{\partial u}{\partial z}\right) + v\left(\frac{\partial w}{\partial y} - \frac{\partial v}{\partial z}\right).$$

The right-hand side has obvious symmetry. Let' multiply the first of (41) by u the second by v and the third by w and add the resulting equality.

We state that all terms of the right-hand sides cancel each other and the resulting equation takes the form

$$u\frac{\partial(d - \alpha_4 - \beta_4 - \gamma_4)}{\partial x} + v\frac{\partial(d - \alpha_4 - \beta_4 - \gamma_4)}{\partial y} +$$

$$w\frac{\partial(d - \alpha_4 - \beta_4 - \gamma_4)}{\partial z} = 0. \qquad (42)$$

Equation of (42) can be written in vector form as

$$\vec{U} \cdot \vec{\nabla}(d - \alpha_4 - \beta_4 - \gamma_4) = 0. \qquad (43)$$

The latter implies that the value of $(d - \alpha_4 - \beta_4 - \gamma_4)$ does not change on direction given by the velocity vector $\vec{U}(u, v, w)$. It means that it does not change along the stream line. So, along the stream line the following equality hold

$$(d - \alpha_4 - \beta_4 - \gamma_4) = -C_{sl}, \qquad (44)$$

where C_{sl} is a constant value to the selected stream line.

The value of d determined by (44) is $d = \alpha_4 + \beta_4 + \gamma_4 - C_{sl}$. Substituting this in (39), we obtain just along the stream line equality as the next

$$p + \Phi + \frac{U^2}{2} - C_{sl} = 0. \qquad (45)$$

This is equivalent to Bernoulli's integral (6). Thus, the Bernoulli's integral (6) is a special case of the root integral (10-18).

Proof of Theorem **3.2** *is complete.*

3.3. Integral of Euler — Bernoulli as Special Case of the Root Integral

Theorem 3.3. Integral of Euler — Bernoulli (8) is a special case of the root integral (10-18).

Proof of the Theorem 3.3. We take the assumptions as a basis ones for derivation of Euler — Bernoulli integral [1]. There are three of them. The medium is assumed an ideal, motion is steady-state and in addition to this the equality (7) must be fulfilled. We show that under these conditions

Integrals of the Navier – Stokes and Euler Equations ...

from (10-18) should be (8).

Assumptions about the ideality of the medium and the steady-state nature of motion coincides with the assumptions of section **3.2** It means that we can use some relations of the previous section. In particular hold the equations (39) and (41). But subject to (7) each of the right-hand sides of (41) vanishes. Thus, in our case hold the next

$$\frac{\partial(d - \alpha_4 - \beta_4 - \gamma_4)}{\partial x} = \frac{\partial(d - \alpha_4 - \beta_4 - \gamma_4)}{\partial y} = \frac{\partial(d - \alpha_4 - \beta_4 - \gamma_4)}{\partial z} = 0. \tag{46}$$

This implies that value of the $(d - \alpha_4 - \beta_4 - \gamma_4)$ is independent of spatial coordinates and it equal to absolute constant. We introduce the notation for it according to

$$(d - \alpha_4 - \beta_4 - \gamma_4) = -C. \tag{47}$$

Substituting (47) in (39) we get

$$p + \Phi + \frac{U^2}{2} - C = 0. \tag{48}$$

The latter is equivalent to (8). Thus, Euler — Bernoulli integral is the special case of integral (10-18).

Proof of Theorem **3.3** *is complete.*

3.4. Tree of Integrals for Motion of Incompressible Medium

Thus, the integrals (5), (6), (8) are special cases of the root integral represented by nine ratios (10-18). We say that (10-18) is integral "A". The significance of this integral is that it can be a basis for other new integrals. Let's point out some of these cases.

We consider the integral for motion of an ideal incompressible medium as the first. We call it as integral "I". It is derived from "A" as a result of simplifications due to $\frac{1}{Re} = 0$. In this case changes will occur only on five ratios (11-15). As a result, fall out of consideration of the group terms are proportional to $\frac{1}{Re}$. The resulting integral "I" presented by nine ratios as well as the root integral "A".

For the steady - state motion of incompressible viscous medium occurs integral, which we call as integral "S". It's derived from "A" as a result of simplification $\frac{\partial}{\partial t} = 0$. In this case three relations (16-18) has nothing to do with the first six ones and disappears. They replaced with the continuity equation (4). For this case integral will be presented by six relationships. Note that number of associated unknown ψ_i was reduced to six.

Another possible integral in this family corresponds to the case of steady - state motion of an ideal incompressible medium. We call it as "IS" integral. There are two different ways to derive this integral. First one is to use "S" with simplification of $\frac{1}{Re} = 0$. The second one is to use integral "I" with simplification of $\frac{\partial}{\partial t} = 0$. As a result "IS" integral is represented by six relationships. When comparing with the integral "S" changes will take place only in the last five relationships. The left-hand sides of the group terms proportional to $\frac{1}{Re} = 0$ fall out.

There are other, less obvious integrals that are generated by root integral "A". For this we consider the Lagrange — Cauchy integral (5). It is a special case of integral "A". As follows from the above evidence, the integral (5) is the result of changes in certain conditions only one of the nine ratios (10-18). This is the ratio of (10). However, the remaining ratio (11-18) also remain in force. The combination of these eight interconnected relations can be regarded as a new integral. This new integral we call as integral "L2", while the integral (5) we call as integral "L1". So, integral "L2" obtained under the same assumptions that integral "L1" and is given by eight ratios.

A similar situation occurs with the integrals (6) and (8). Both integral correspond to the case of steady - state motion of an ideal medium flow. Both can be obtained from integral "IS", represented by six ratios. From the evidence given in sections **3.2** and **3.3,** implies that both are the result of conversions of ratio (10). However, with the remaining five ratios are also valid. If we denote the Bernoulli's integral (6), as "B1", the Euler — Bernoulli integral (8) as the"E1", that naturally arises the integrals "B2" and "E2". Each of these new integrals represented by five ratios. Moreover "B2" corresponds to the assumptions of the Bernoulli's integral and "E2" corresponds to Euler — Bernoulli one.

Thus, the integral "A" generates six new integrals. These integrals are the next: "S", "I", "IS", "B2", "E2", "L2".

To the various relationships between the integrals would be identified more clearly they are conveniently located as a tree. It follows from the considerations above, that in the base of the tree should be placed integral "A", as the general one. All other integrals are it special cases. They constitute two main branches. The right-hand branch are the integrals of motion for viscous medium "S", "L1", "L2". The left-hand branch are integrals for the particular cases of an ideal medium "I", "IS", "B1", "B2", "E1", "E2". All of these ten integrals form a tree of integrals for motion of an incompressible medium. It is represented on Figure 2.

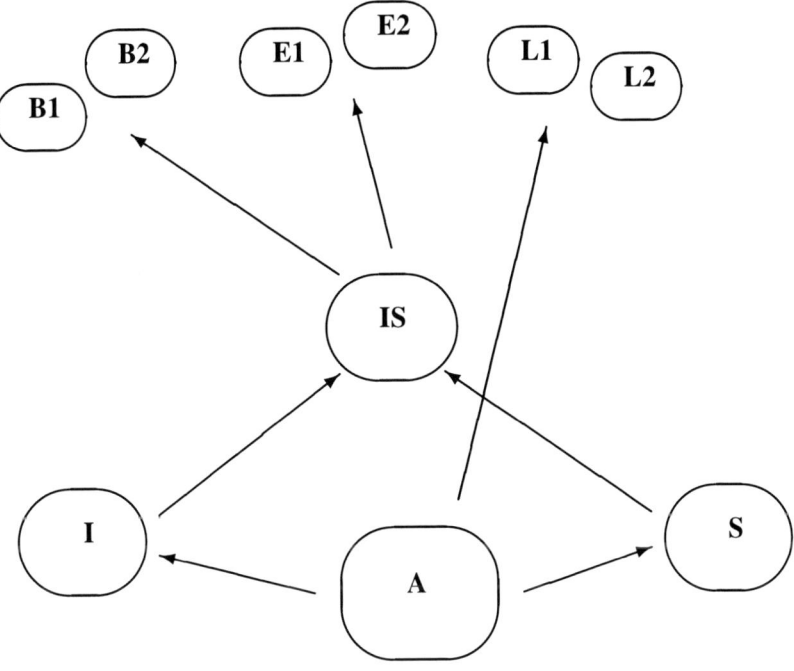

Figure 2. Tree of integrals for motion of an incompressible medium.

4. DISCUSSION

Sometimes in the works on fluid mechanics they use the simplified name for the classical integrals (5), (6), (8) calling each of them the Bernoulli integral [6-7] on the grounds that each of them contains the Bernoulli's trinomial. The author believes this approach is not correct. Each of these integrals has its own specifics, obtained under strictly defined assumptions and has its own scope. Therefore, these integrals must be called differently.

The same is true with respect to the new integrals presented in Figure 2. Some of them contains the Bernoulli's trinomial, but this is where the similarities end and there are clearly more significant differences. Therefore, each needs its own designation and its name.

CONCLUSION

The performed research made it possible to obtain new integrals of motion of an incompressible medium and to trace all possible relationships between them and the known classical integrals. Integral "A" with good reason can be called the root integral of motion of an incompressible medium. All other considered integrals are its special cases. Each integral are relations between the defining values. These relations can simplify the solution of problems, leading them to more simple tasks. This approach is implemented by author in the study of classical problems of fluid mechanics [13-14], in solving purely practical problems [15-16] and in construction of new exact solutions of the Navier-Stokes and the Euler equations [17-20].

REFERENCES

[1] Sedov L.I., Continuum mechanics, Part 2, *Nauka, Moscow, Russia*, 1970, (in Russian).

[2] Loitsaynskiy L.G., Mechanics of Fluid and Gas, *Nauka, Moscow, Russia*, 1987, (in Russian).

[3] Lodizhenskaya O.A., *The Mathematical Theory of Viscous Incompressible Fluid*, Gordon and Breach, NY, USA, 1969.

[4] Fefferman Charles L., Existence and Smoothness of the Navier — Stokes Equation, *Preprint, Princeton Univ., Math. Dept., Princeton, NJ, USA*, 2000, pp. 1-5.

[5] Dynnikova G.Y., An analogue of the Bernoulli and Cauchy-Lagrange integrals for the unsteady vortex flow of an ideal incompressible fluid, *Izvesyia RAS, J. Mech. Fluid and Gas*, **2000**, 1, P. 31-41, (in Russian).

[6] Ershkov S.V., On existence of general solution of the Navier — Stokes equations for 3D non-stationary incompressible flow, *Int. J. Fluid Mech. Research*, **2015**, 06, 42(3), P. 206-215.

[7] Ershkov S.V., Non-stationary creeping flows for incompressible 3D Navier — Stokes equations, *Europ. J. Mech., B/Fluids*, **2017**, 61(1), P. 154-159.

[8] Koptev A.V., Integrals of the Navier — Stokes equations, *Trudy Sredn.-Volzh.Mat.Obshch.*, **2004**, 6(1) pp. 215-225, (in Russian).

[9] Koptev A.V., Integrals of Motion of an Incompressible medium Flow. From Classic to Modern. In book: *Handbook on Navier — Stokes Equations. Theory and Applied Analysis* – edition by D. Campos, Nova Science Rublishers, NY, USA 2017, pp.443-460.

[10] Koptev A.V., First integral and ways of further integration of Navier — Stokes equations, *Izvestia RGPU im. Gertsena, Saint-Petersburg*, **2012**, 147, pp. 7-17, (in Russian).

[11] Koptev A.V., Integrals of Motion of an incompressible fluid, *Proc. of 12-th All-Russian Congress on fundamental problems of theoretical and applied mechanics, State Univ. of Bashkortostan, Ufa, Russia, August 2019*, **2019**, pp. 159-162, (in Russian).

[12] Polynin A.D., Zhurov A.I., Parametically defined nonlinear differential equations and their solutions, *Appl. Math. Letters*, **2016**, 55, P. 72-80.

[13] Koptev A.V., D'Alembert paradox in near real conditions, *J. Siberian Federal Univ., Math. and Phys.*, **2017**, 10(2), P. 170-180.

[14] Koptev A.V., Nonlinear effects in Poiseuille problem, *J. Siberian Federal Univ., Math. and Phys.*, **2013**, 6(3), P. 308-314.

[15] Koptev A.V., Dynamic responce of an underwater pipeline on the sea currents, *Vestnik GUMRF im. adm. S.O. Makarova, Saint-Petersburg*, **2014**, 4(26), pp. 107-114, (in Russian).

[16] Koptev A.V., Theoretical research of the flow around the cylinder of an ideal incompressible medium in the presence of a shielding effect, *Vestnik GUMRF im. adm. S.O. Makarova, Saint-Petersburg*, **2016**, 2(36), pp. 127-137, (in Russian).

[17] Koptev A.V., The solution of initial and boundary value problem for 3D Navier — Stokes equations and its features, *Izvestia RGPU im. Gertsena, Saint-Petersburg*, **2014**, 165, pp. 7-18, (in Russian).

[18] Koptev A.V., Generator of solution of 2D Navier — Stokes equations, *J. Siberian Federal Univ., Math. and Phys.*, **2014**, 7(3), P. 324-330.

[19] Koptev A.V., Exact solutions of 3D Navier — Stokes equations, *J. Siberian Federal Univ., Math. and Phys.*, **2020**, 13(3), P. 306-313.

[20] Koptev A.V., Method for solving the Navier — Stokes and Euler equations of motion for incompressible media, *J. Mathematical Sciences.*, Springer, NY, USA, **2020**, 250(1), P. 254-265.

Chapter 4

DEEP WATER MOVEMENT

Alexander V. Koptev[*]

Math. Dept., Admiral Makarov State University
of Maritime and Inland Shipping, Saint-Petersburg, Russia

Abstract

We consider the movement in a deep water under action of gravity based on 3D Navier – Stokes equations for an incompressible medium. The main unknowns are pressure and three components of the velocity vector, while each of these quantities is a function of spatial coordinates and time. The density and kinematic viscosity of the fluid are assumed to be constant. We neglect an influence of all bounding surfaces with the exception of the free surface. Along the free surface we set the pressure constancy condition. The integral of 3D Navier – Stokes equations and the generator of solutions obtained by the author are proposed as the initial relations. As a result of the application of the methods of partial differential equations and mathematical physics the exact solutions of the Navier – Stokes equations characterizing the movement in a deep water and near the free surface are found out. These results make possible to carry out a study of the nonlinear effects inherent in gravitational waves on the water surface. In particular we obtained the equations for free surface profile depending on coordinates, time and other governing parameters.

---

[*] Corresponding Author's Email: Alex.Koptev@mail.ru.

Keywords: movement, deep water, viscous fluid, Navier - Stokes equations, gravity, velocity, pressure, integral, free surface profile

1. INTRODUCTION

The movement of water masses at a depth determine many factors among which one can distinguish climatic and weather changes, intensity of wave formation, navigation conditions and safety of navigation. The study of water masses movement at depth is an important task, and the results have applications in oceanology, climatology, shipbuilding and in other areas.

In this paper we propose a theoretical study of movement in a deep water under action of gravity. The research is based on 3D Navier — Stokes equations for isothermal motion of incompressible fluid.

We choose a coordinate system so, that the OZ axis is directed vertically upwards and the XOY plane coincides with the horizon plane. For the case when gravity play the role of an external mass force the dimensionless form of the Navier — Stokes equations and the equation of continuity can be represented as

$$\frac{\partial u}{\partial t} + u\frac{\partial u}{\partial x} + v\frac{\partial u}{\partial y} + w\frac{\partial u}{\partial z} = -\frac{\partial p}{\partial x} + \frac{1}{Re}\Delta u, \qquad (1)$$

$$\frac{\partial v}{\partial t} + u\frac{\partial v}{\partial x} + v\frac{\partial v}{\partial y} + w\frac{\partial v}{\partial z} = -\frac{\partial p}{\partial y} + \frac{1}{Re}\Delta v, \qquad (2)$$

$$\frac{\partial w}{\partial t} + u\frac{\partial w}{\partial x} + v\frac{\partial w}{\partial y} + w\frac{\partial w}{\partial z} = -g - \frac{\partial p}{\partial z} + \frac{1}{Re}\Delta w, \qquad (3)$$

$$\frac{\partial u}{\partial x} + \frac{\partial v}{\partial y} + \frac{\partial w}{\partial z} = 0, \qquad (4)$$

where the components of the velocity vector u, v, w and the pressure p are main unknowns and each of them depend on spatial coordinates x, y, z and time t; Δ denote 3D Laplace operator on coordinates,

$$\Delta = \frac{\partial^2}{\partial x^2} + \frac{\partial^2}{\partial y^2} + \frac{\partial^2}{\partial z^2};$$

g is the acceleration due to gravity; Re is a dimensionless non-negative parameter called the Reynolds number.

We will assume that the fluid flow domain is not limited by anything, both in the direction of the OX axis and in the direction of the OY axis, while in the direction of the OZ axis it is limited by the free surface. We assume that along the free surface the pressure is constant [1-2].

Thus, the problem is reduced to solving the equations (1-4) under condition of pressure constant along the free surface.

The problem of study and resolution of the Navier - Stokes equations is one of the important problems of modern mathematics [3-4]. It should be admitted that today there are attempts but is no generally accepted approach to resolution of 3D Navier – Stokes equations [5-7]. On the paper under consideration we propose new approach to this problem.

2. METHODS

2.1. First Integral

To solve the problem it is proposed to proceed not from equations (1-4) directly, but from the first integral of these equations previously obtained by the author [8-9]. The integral of equations (1-4) is reduced to nine relations connecting the main unknowns u, v, w, p, associated unknowns Ψ_i and an arbitrary additive functions of three variables $\alpha_j, \beta_j, \gamma_j, \delta_j$. The relations representing the first integral in the simplest notation have the form

$$p - p_0 + gz + \frac{U^2}{2} + d + d_t = \alpha_4 + \beta_4 + \gamma_4, \qquad (5)$$

$$u^2 - v^2 + \frac{2}{Re}(-\frac{\partial u}{\partial x} + \frac{\partial v}{\partial y}) = -\frac{\partial^2 \Psi_{10}}{\partial x^2} + \frac{\partial^2 \Psi_{10}}{\partial y^2} - \frac{\partial^2 \Psi_{11}}{\partial z^2} - \frac{\partial^2 \Psi_{12}}{\partial z^2} + \frac{\partial^2 \Psi_{15}}{\partial y \partial z} +$$

$$\frac{\partial^2 \Psi_{14}}{\partial x \partial z} + \frac{\partial}{\partial t}(-\frac{\partial \Psi_1}{\partial x} + \frac{\partial \Psi_3}{\partial y} + \frac{\partial(\Psi_5 + \Psi_6)}{\partial z}) + 3(\alpha_4 - \beta_4), \qquad (6)$$

$$v^2 - w^2 + \frac{2}{Re}(-\frac{\partial v}{\partial y} + \frac{\partial w}{\partial z}) = \frac{\partial^2 \Psi_{10}}{\partial x^2} + \frac{\partial^2 \Psi_{11}}{\partial x^2} - \frac{\partial^2 \Psi_{12}}{\partial y^2} + \frac{\partial^2 \Psi_{12}}{\partial z^2} - \frac{\partial^2 \Psi_{13}}{\partial x \partial y} -$$

$$\frac{\partial^2 \Psi_{14}}{\partial x \partial z} + \frac{\partial}{\partial t}(\frac{\partial(\Psi_1 + \Psi_2)}{\partial x} + \frac{\partial \Psi_4}{\partial y} - \frac{\partial \Psi_6}{\partial z}) + 3(\beta_4 - \gamma_4), \qquad (7)$$

$$uv - \frac{1}{Re}(\frac{\partial v}{\partial x} + \frac{\partial u}{\partial y}) = -\frac{\partial^2 \Psi_{10}}{\partial x \partial y} + \frac{1}{2}\frac{\partial}{\partial z}(-\frac{\partial \Psi_{15}}{\partial x} + \frac{\partial \Psi_{14}}{\partial y} + \frac{\partial \Psi_{13}}{\partial z}) +$$

$$\frac{1}{2}\frac{\partial}{\partial t}(-\frac{\partial \Psi_3}{\partial x} - \frac{\partial \Psi_1}{\partial y} - \frac{\partial(\Psi_8 + \Psi_9)}{\partial z}) + \frac{1}{2}(-\frac{\partial \alpha_1}{\partial z} - \frac{\partial \alpha_3}{\partial t} + \frac{\partial \beta_1}{\partial z} - \frac{\partial \beta_2}{\partial t}), \quad (8)$$

$$uw - \frac{1}{Re}(\frac{\partial w}{\partial x} + \frac{\partial u}{\partial z}) = \frac{\partial^2 \Psi_{11}}{\partial x \partial z} - \frac{1}{2}\frac{\partial}{\partial y}(\frac{\partial \Psi_{15}}{\partial x} + \frac{\partial \Psi_{14}}{\partial y} + \frac{\partial \Psi_{13}}{\partial z}) +$$

$$\frac{1}{2}\frac{\partial}{\partial t}(-\frac{\partial \Psi_5}{\partial x} + \frac{\partial(\Psi_9 - \Psi_7)}{\partial y} + \frac{\partial \Psi_2}{\partial z}) + \frac{1}{2}(-\frac{\partial \alpha_1}{\partial y} - \frac{\partial \alpha_2}{\partial t} + \frac{\partial \gamma_1}{\partial y} - \frac{\partial \gamma_3}{\partial t}), \quad (9)$$

$$vw - \frac{1}{Re}(\frac{\partial w}{\partial y} + \frac{\partial v}{\partial z}) = -\frac{\partial^2 \Psi_{12}}{\partial y \partial z} + \frac{1}{2}\frac{\partial}{\partial x}(\frac{\partial \Psi_{14}}{\partial y} + \frac{\partial \Psi_{15}}{\partial x} - \frac{\partial \Psi_{13}}{\partial z}) +$$

$$\frac{1}{2}\frac{\partial}{\partial t}(\frac{\partial(\Psi_7 + \Psi_8)}{\partial x} + \frac{\partial \Psi_6}{\partial y} + \frac{\partial \Psi_4}{\partial z}) - \frac{1}{2}(\frac{\partial \beta_1}{\partial x} + \frac{\partial \beta_2}{\partial t} + \frac{\partial \gamma_1}{\partial x} + \frac{\partial \gamma_2}{\partial t}), \quad (10)$$

$$u = \frac{1}{2}(\frac{\partial}{\partial y}(-\frac{\partial \Psi_3}{\partial x} + \frac{\partial \Psi_1}{\partial y} + \frac{\partial \Psi_7}{\partial z}) + \frac{\partial}{\partial z}(-\frac{\partial \Psi_5}{\partial x} + \frac{\partial \Psi_8}{\partial y} - \frac{\partial \Psi_2}{\partial z})) +$$

$$\frac{1}{2}(\frac{\partial \alpha_2}{\partial z} + \frac{\partial \alpha_3}{\partial y} + \frac{\partial \delta_1}{\partial y} + \frac{\partial \delta_2}{\partial z}), \quad (11)$$

$$v = \frac{1}{2}(\frac{\partial}{\partial x}(\frac{\partial \Psi_3}{\partial x} - \frac{\partial \Psi_1}{\partial y} - \frac{\partial \Psi_7}{\partial z}) + \frac{\partial}{\partial z}(\frac{\partial \Psi_9}{\partial x} + \frac{\partial \Psi_6}{\partial y} - \frac{\partial \Psi_4}{\partial z})) +$$

$$\frac{1}{2}(\frac{\partial \beta_2}{\partial x} + \frac{\partial \beta_3}{\partial z} - \frac{\partial \delta_1}{\partial x} + \frac{\partial \delta_3}{\partial z}), \quad (12)$$

$$w = \frac{1}{2}(\frac{\partial}{\partial x}(\frac{\partial \Psi_5}{\partial x} - \frac{\partial \Psi_8}{\partial y} + \frac{\partial \Psi_2}{\partial z}) + \frac{\partial}{\partial y}(-\frac{\partial \Psi_9}{\partial x} - \frac{\partial \Psi_6}{\partial y} + \frac{\partial \Psi_4}{\partial z})) +$$

$$\frac{1}{2}(\frac{\partial \gamma_2}{\partial y} + \frac{\partial \gamma_3}{\partial x} - \frac{\partial \delta_2}{\partial x} - \frac{\partial \delta_3}{\partial y}). \quad (13)$$

Functions Ψ_i where $i = 1, 2, ..., 15$ denotes new associated unknowns arising in the process of integration. For the case under consideration there are fifteen and they complicate the system of unknowns. In the papers [8-9] we introduce for them name "stream pseudo-functions". Thus, there are nineteen unknowns in a total, namely four main unknowns and fifteen associated ones.

The ratio (5) contain an additional values p_0, $\frac{U^2}{2}$, d and d_t. The first one is the additive pressure constant, the second one is the dimensionless velocity head

$$\frac{U^2}{2} = \frac{u^2 + v^2 + w^2}{2}.$$

Values d and d_t are dissipative terms defined as

$$d = -\frac{U^2}{6} - \frac{1}{3}(\Delta_{xy}\Psi_{10} - \Delta_{xz}\Psi_{11} + \Delta_{yz}\Psi_{12} + \frac{\partial^2\Psi_{13}}{\partial x \partial y} - \frac{\partial^2\Psi_{14}}{\partial x \partial z} + \frac{\partial^2\Psi_{15}}{\partial y \partial z}), \quad (14)$$

$$d_t = \frac{1}{3}\frac{\partial}{\partial t}\left(\frac{\partial(\Psi_2 - \Psi_1)}{\partial x} + \frac{\partial(\Psi_4 - \Psi_3)}{\partial y} + \frac{\partial(\Psi_6 - \Psi_5)}{\partial z}\right). \quad (15)$$

Symbols Δ_{yz}, Δ_{xz}, Δ_{xy} in (14) denote the incomplete Laplace operators with respect to spatial coordinates

$$\Delta_{yz} = \frac{\partial^2}{\partial y^2} + \frac{\partial^2}{\partial z^2}, \quad \Delta_{xz} = \frac{\partial^2}{\partial x^2} + \frac{\partial^2}{\partial z^2}, \quad \Delta_{xy} = \frac{\partial^2}{\partial x^2} + \frac{\partial^2}{\partial y^2}.$$

The order of derivatives with respect to main unknowns in (5 - 13) is one less than their order in the original equations (1 - 4). When considering together the ratios (5 - 13) represent first integral of 3D Navier – Stokes equations (1 - 4).

Existence of integral allow to go forward to solution of the problem in a new way. A favorable circumstance is the excess of the number of unknowns over the number of equations since it becomes possible to satisfy some additional conditions.

2.2. Generator of Solutions

Let us introduce the concept of a generator of solutions to equations (1 - 4). By this we mean relations that allow to construct the solutions of (1 - 4) in a quick way. The generator of solutions for 2D Navier-Stokes equations was previously proposed by the author [10]. Let's set the goal of building a generator of solutions for 3D equations.

As the initial step we briefly analyze the ratios (5 - 13) representing the first integral of 3D Navier – Stokes equations. Let's pay special attention to ratios (5) and (11 - 13). They give the expressions for the main unknowns u, v, w, p in terms of associated unknowns Ψ_i. It is fair to conclude that these four ratios determine the general structure of solutions to equations (1 - 4).

Ratio (5) is the special one, since only this contain the unknown p. In the way of practical resolution of equations this ratio should be used at the last stage when all other unknowns have already been found.

When considering the ratios (6 - 10) in general the following regularities attract attention. In the right-hand sides of (11 - 13) there are derivatives of only the first nine associated unknowns Ψ_k, $k = 1, 2, ...9$, while unknowns of this type in total are fifteen. Unknowns Ψ_k when $k = 10, 11, ..., 15$ do not appear in ratios (11 - 13) while in ratios (6 - 10) these unknowns are present as a linear combination of second derivatives. It is possible to exclude these unknowns from (6 - 7) and obtain general relations ensuring the solvability of all the other ones. The procedure for constructing such relations is described below.

Let's consider the simplest case when the additive functions α_i, β_i, γ_i, δ_i equal zero. We denote sums of all terms of the equations, respectively (6 - 10) independent of unknowns Ψ_k ($k = 1, 2, ..., 9$), as f_j ($j = 2, 3, ..., 6$). So, we obtain

$$u^2 - v^2 + \frac{2}{Re}\left(-\frac{\partial u}{\partial x} + \frac{\partial v}{\partial y}\right) + \frac{\partial}{\partial t}\left(\frac{\partial \Psi_1}{\partial x} - \frac{\partial \Psi_3}{\partial y} - \frac{\partial \Psi_5}{\partial z} - \frac{\partial \Psi_6}{\partial z}\right) = f_2,$$

$$v^2 - w^2 + \frac{2}{Re}\left(-\frac{\partial v}{\partial y} + \frac{\partial w}{\partial z}\right) - \frac{\partial}{\partial t}\left(\frac{\partial \Psi_1}{\partial x} + \frac{\partial \Psi_2}{\partial x} + \frac{\partial \Psi_4}{\partial y} - \frac{\partial \Psi_6}{\partial z}\right) = f_3,$$

$$uv - \frac{1}{Re}\left(\frac{\partial v}{\partial x} + \frac{\partial u}{\partial y}\right) + \frac{1}{2}\frac{\partial}{\partial t}\left(\frac{\partial \Psi_3}{\partial x} + \frac{\partial \Psi_1}{\partial y} + \frac{\partial \Psi_8}{\partial z} + \frac{\partial \Psi_9}{\partial z}\right) = f_4, \quad (16)$$

$$uw - \frac{1}{Re}\left(\frac{\partial w}{\partial x} + \frac{\partial u}{\partial z}\right) + \frac{1}{2}\frac{\partial}{\partial t}\left(\frac{\partial \Psi_5}{\partial x} + \frac{\partial \Psi_7}{\partial y} - \frac{\partial \Psi_9}{\partial y} - \frac{\partial \Psi_2}{\partial z}\right) = f_5,$$

$$vw - \frac{1}{Re}\left(\frac{\partial w}{\partial y} + \frac{\partial v}{\partial z}\right) - \frac{1}{2}\frac{\partial}{\partial t}\left(\frac{\partial \Psi_7}{\partial x} + \frac{\partial \Psi_8}{\partial x} + \frac{\partial \Psi_6}{\partial y} + \frac{\partial \Psi_4}{\partial z}\right) = f_6,$$

As a result, five nonlinear equations (6 - 10) we can represent in the form of

$$f_2 = -\frac{\partial^2 \Psi_{10}}{\partial x^2} + \frac{\partial^2 \Psi_{10}}{\partial y^2} - \frac{\partial^2 \Psi_{11}}{\partial z^2} - \frac{\partial^2 \Psi_{12}}{\partial z^2} + \frac{\partial^2 \Psi_{15}}{\partial y \partial z} + \frac{\partial^2 \Psi_{14}}{\partial x \partial z}, \quad (17)$$

$$f_3 = \frac{\partial^2 \Psi_{10}}{\partial x^2} + \frac{\partial^2 \Psi_{11}}{\partial x^2} - \frac{\partial^2 \Psi_{12}}{\partial y^2} + \frac{\partial^2 \Psi_{12}}{\partial z^2} - \frac{\partial^2 \Psi_{13}}{\partial x \partial y} - \frac{\partial^2 \Psi_{14}}{\partial x \partial z}, \quad (18)$$

$$f_4 = -\frac{\partial^2 \Psi_{10}}{\partial x \partial y} + \frac{1}{2}\frac{\partial}{\partial z}\left(-\frac{\partial \Psi_{15}}{\partial x} + \frac{\partial \Psi_{14}}{\partial y} + \frac{\partial \Psi_{13}}{\partial z}\right), \qquad (19)$$

$$f_5 = \frac{\partial^2 \Psi_{11}}{\partial x \partial z} - \frac{1}{2}\frac{\partial}{\partial y}\left(\frac{\partial \Psi_{15}}{\partial x} + \frac{\partial \Psi_{14}}{\partial y} + \frac{\partial \Psi_{13}}{\partial z}\right), \qquad (20)$$

$$f_6 = -\frac{\partial^2 \Psi_{12}}{\partial y \partial z} + \frac{1}{2}\frac{\partial}{\partial x}\left(\frac{\partial \Psi_{14}}{\partial y} + \frac{\partial \Psi_{15}}{\partial x} - \frac{\partial \Psi_{13}}{\partial z}\right). \qquad (21)$$

Let's eliminate from equations (17-18) terms with unknowns Ψ_k for $k = 10, 11, ..., 15$ with help of (19-21). To do this we calculate the derivatives of (17 - 21) with respect to coordinates term by term and select the necessary linear combinations to exclude terms with the specified unknowns. We divide the chain of transformations into two stages.

Stage 1. We take $\frac{\partial}{\partial y}$ from (17), $-\frac{\partial}{\partial x}$ from (19), $-\frac{\partial}{\partial z}$ from (21). Adding the results we obtain

$$\frac{\partial f_2}{\partial y} - \frac{\partial f_4}{\partial x} - \frac{\partial f_6}{\partial z} = \frac{\partial^3 \Psi_{10}}{\partial y^3} - \frac{\partial^3 \Psi_{11}}{\partial y \partial z^2} + \frac{\partial^3 \Psi_{15}}{\partial y^2 \partial z}. \qquad (22)$$

We take $\frac{\partial}{\partial x}$ from (22), $\frac{\partial^2}{\partial y^2}$ from (19), $\frac{\partial^2}{\partial y \partial z}$ from (20). Further, adding up the results, we come to

$$\frac{\partial^2 f_2}{\partial x \partial y} - \frac{\partial^2 f_4}{\partial x^2} + \frac{\partial^2 f_4}{\partial y^2} + \frac{\partial^2 f_5}{\partial y \partial z} - \frac{\partial^2 f_6}{\partial x \partial z} = 0. \qquad (23)$$

Stage 2. We take $\frac{\partial}{\partial y}$ from (18), $\frac{\partial}{\partial x}$ from (19), $\frac{\partial}{\partial z}$ from (21). Adding the results we obtain

$$\frac{\partial f_3}{\partial y} + \frac{\partial f_4}{\partial x} + \frac{\partial f_6}{\partial z} = \frac{\partial^3 \Psi_{11}}{\partial x^2 \partial y} - \frac{\partial^3 \Psi_{12}}{\partial y^3} - \frac{\partial^3 \Psi_{13}}{\partial x \partial y^2}. \qquad (24)$$

We take $\frac{\partial}{\partial z}$ from (24), $-\frac{\partial^2}{\partial x \partial y}$ from (20), $-\frac{\partial^2}{\partial y^2}$ from (21). Adding the results we arrive to

$$\frac{\partial^2 f_3}{\partial y \partial z} + \frac{\partial^2 f_4}{\partial x \partial z} - \frac{\partial^2 f_5}{\partial x \partial y} - \frac{\partial^2 f_6}{\partial y^2} + \frac{\partial^2 f_6}{\partial z^2} = 0. \qquad (25)$$

It becomes clear on the basis of (16) that only nine unknowns are present in equations (23) and (25). These unknowns are Ψ_k where $k =$

1, 2, ..., 9. This fact is obvious since according to (11 - 13) u, v, w are defined in terms of these unknowns only. Any set of functions $\Psi_1, \Psi_2, ..., \Psi_9$ that satisfy this system allow to determine all other unknowns including the main ones. Firstly are found u, v, w according to (11 - 13). Then by using of (16) are defined f_j for $j = 2, 3, ..., 6$. Further with help of (17 - 21) we determine six unknowns $\Psi_{10}, \Psi_{11}, ..., \Psi_{15}$. And at the last step, using the ratios (5) and (14 - 15), we determine the unknown p.

Hence, the system of equations (23) and (25) can be considered as the generator of solutions to $3D$ Navier — Stokes equations (1 - 4).

2.3. Exact Solutions Describing Deep Water Movement

Let's start with the formulas (11 - 13) that define general structure for unknowns u, v, w. We confine ourselves to the simplest case of an analytical solutions, assuming the next

$$\frac{\partial \Psi_1}{\partial y} - \frac{\partial \Psi_3}{\partial x} + \frac{\partial \Psi_7}{\partial z} = A(t)e^{(n_1 x + m_1 y + l_1 z)},$$

$$-\frac{\partial \Psi_5}{\partial x} + \frac{\partial \Psi_8}{\partial y} - \frac{\partial \Psi_2}{\partial z} = B(t)e^{(n_2 x + m_2 y + l_2 z)}, \qquad (26)$$

$$\frac{\partial \Psi_9}{\partial x} + \frac{\partial \Psi_6}{\partial y} - \frac{\partial \Psi_4}{\partial z} = C(t)e^{(n_3 x + m_3 y + l_3 z)},$$

where wave numbers $n_k, m_k, l_k, (k = 1, 2, 3)$ are some nonzero constants, then how $A(t), B(t), C(t)$ are some functions of time.

Suppose that unknowns Ψ_i $(i = 1, 2, ..., 9)$, existing in (26), are defined as

$$\Psi_1 = \frac{a_1(t)}{m_1} e^{(n_1 x + m_1 y + l_1 z)}, \quad \Psi_3 = \frac{a_3(t)}{n_1} e^{(n_1 x + m_1 y + l_1 z)},$$

$$\Psi_7 = \frac{a_7(t)}{l_1} e^{(n_1 x + m_1 y + l_1 z)},$$

$$\Psi_2 = \frac{b_2(t)}{l_2} e^{(n_2 x + m_2 y + l_2 z)}, \quad \Psi_5 = \frac{b_5(t)}{n_2} e^{(n_2 x + m_2 y + l_2 z)},$$

$$\Psi_8 = \frac{b_8(t)}{m_2} e^{(n_2 x + m_2 y + l_2 z)}, \qquad (27)$$

$$\Psi_4 = \frac{c_4(t)}{l_3} e^{(n_3 x + m_3 y + l_3 z)}, \quad \Psi_6 = \frac{c_6(t)}{m_3} e^{(n_3 x + m_3 y + l_3 z)},$$

$$\Psi_9 = \frac{c_9(t)}{n_3} e^{(n_3 x + m_3 y + l_3 z)}.$$

We also assume that the amplitude functions satisfy relations as the next

$$a_1(t) - a_3(t) + a_7(t) = A(t), \quad -b_2(t) - b_5(t) + b_8(t) = B(t),$$

$$-c_4(t) + c_6(t) + c_9(t) = C(t). \tag{28}$$

As a result the relations (26) are carried out identically and for the main unknowns u, v, w due to (11-13) we arrive to expressions as follows

$$u = \frac{1}{2}(Am_1 e^{(n_1 x + m_1 y + l_1 z)} + Bl_2 e^{(n_2 x + m_2 y + l_2 z)}),$$

$$v = \frac{1}{2}(-An_1 e^{(n_1 x + m_1 y + l_1 z)} + Cl_3 e^{(n_3 x + m_3 y + l_3 z)}), \tag{29}$$

$$w = \frac{1}{2}(-Bn_2 e^{(n_2 x + m_2 y + l_2 z)} - Cm_3 e^{(n_3 x + m_3 y + l_3 z)}).$$

Thus, we look for solutions u, v, w in form (29), where n_k, m_k, l_k for $k = 1, 2, 3$ are yet now uncertain nonzero wave numbers and $A(t), B(t), C(t)$ are uncertain functions of time.

Let's refer to generator of solutions (23), (25) and find the restrictions under which these equations are obviously done.

Substitution of (16) into (23), (25) with help of (26) and (29) leads to following results. Components of two kinds will present on (23), (25). First kind is linear combinations of quantities $e^{(n_1 x + m_1 y + l_1 z)}$, $e^{(n_2 x + m_2 y + l_2 z)}$, $e^{(n_3 x + m_3 y + l_3 z)}$. The second kind of components is quadratic combinations of these quantities. Members of the first kind are mutually canceled if functions $A(t), B(t), C(t)$ satisfy the ordinary differential equations of the first order

$$\frac{dA}{dt} = \frac{A}{Re}(n_1^2 + m_1^2 + l_1^2), \quad \frac{dB}{dt} = \frac{B}{Re}(n_2^2 + m_2^2 + l_2^2),$$

$$\frac{dC}{dt} = \frac{C}{Re}(n_3^2 + m_3^2 + l_3^2). \tag{30}$$

Members of the second kind are also mutually canceled and equations (23), (25) are performed identically if wave numbers n_k, m_k, l_k satisfy the system of six algebraic equations

$$2m_1 l_2(n_1 + n_2)(m_1 + m_2) + n_1 l_2(n_1 + n_2)^2 - n_1 l_2(m_1 + m_2)^2 -$$
$$m_1 n_2(m_1 + m_2)(l_1 + l_2) - n_1 n_2(n_1 + n_2)(l_1 + l_2) = 0, \qquad (31)$$

$$-n_1 l_2(n_1 + n_2)(l_1 + l_2) + m_1 n_2(n_1 + n_2)(m_1 + m_2) - n_1 n_2(m_1 + m_2)^2 +$$
$$n_1 n_2(l_1 + l_2)^2 = 0, \qquad (32)$$

$$-2n_2 m_3(m_2 + m_3)(l_2 + l_3) + n_2 l_3(m_2 + m_3)^2 - n_2 l_3(l_2 + l_3)^2 +$$
$$l_2 l_3(n_2 + n_3)(l_2 + l_3) + m_3 l_2(n_2 + n_3)(m_2 + m_3) = 0, \qquad (33)$$

$$-m_3 l_2(m_2 + m_3)(l_2 + l_3) + n_2 l_3(n_2 + n_3)(l_2 + l_3) - l_2 l_3(n_2 + n_3)^2 +$$
$$l_2 l_3(m_2 + m_3)^2 = 0, \qquad (34)$$

$$2n_1 l_3(n_1 + n_3)(m_1 + m_3) - m_1 l_3(n_1 + n_3)^2 + m_1 l_3(m_1 + m_3)^2 -$$
$$m_1 m_3(m_1 + m_3)(l_1 + l_3) - n_1 m_3(n_1 + n_3)(l_1 + l_3) = 0, \qquad (35)$$

$$-2n_1 l_3(m_1 + m_3)(l_1 + l_3) + m_1 l_3(n_1 + n_3)(l_1 + l_3) + m_1 m_3(n_1 + n_3)(m_1 + m_3) -$$
$$n_1 m_3(m_1 + m_3)^2 + n_1 m_3(l_1 + l_3)^2 = 0. \qquad (36)$$

Note that previous ratios give conditions for the existence of a solution in the form of (29) under the constraints

$$n_k \neq 0, \quad m_k \neq 0, \quad l_k \neq 0, \quad k = 1, 2, 3. \qquad (37)$$

If constraints (37) are violated, then the indicated ratios are replaced with simpler ones.

The solutions of equations (30) are easy to find. They are defined as

$$A(t) = A(0)e^{\frac{(n_1^2 + m_1^2 + l_1^2)t}{Re}}, \quad B(t) = B(0)e^{\frac{(n_2^2 + m_2^2 + l_2^2)t}{Re}},$$

$$C(t) = C(0)e^{\frac{(n_3^2+m_3^2+l_3^2)t}{Re}}, \tag{38}$$

where $A(0), B(0), C(0)$ are an arbitrary constants which define values of the amplitude functions at $t = 0$.

As for the system (31-36) the preliminary analysis show that this system present family of six nonlinear equations with respect to nine unknowns $n_k, m_k, l_k, \ (k = 1, 2, 3)$. Thus, to find the unknown wave numbers we face a purely algebraic problem under condition that the number of unknowns exceed the number of equations.

System (31-36) has many different solutions, some of which are real and complex. Each set of numbers, that satisfy (31-36) generate exact solution of the Navier — Stokes equations (1 - 4). In contrast to work [11] we will focus our efforts on purely real solutions and consider in more detail some of them.

3. SOLUTION OPTIONS

3.1. Solution 1

The simplest solution correspond to the case of propagation of three plane waves in a deep water under condition of their wave vectors (n_k, m_k, l_k) are collinear. This case is realized if n_3, m_3, l_3 are an arbitrary nonzero values and additionally the following relations are satisfy $n_1 = \mu n_3$, $m_1 = \mu m_3$, $l_1 = \mu l_3$, $n_2 = \xi n_3$, $m_2 = \xi m_3$, $l_2 = \xi l_3$. In the latter relations μ and ξ are an arbitrary nonzero real values. For this case all six equations (31-36) are executed identically.

According to (29) and (38) we arrive to expressions as the next

$$u = \frac{1}{2}(A(0)\mu m_3 e^{\frac{N_1^2 t}{Re}+\mu(n_3x+m_3y+l_3z)} + B(0)\xi l_3 e^{\frac{N_2^2 t}{Re}+\xi(n_3x+m_3y+l_3z)}),$$

$$v = \frac{1}{2}(-A(0)\mu n_3 e^{\frac{N_1^2 t}{Re}+\mu(n_3x+m_3y+l_3z)} + C(0)l_3 e^{\frac{N_3^2 t}{Re}+(n_3x+m_3y+l_3z)}), \tag{39}$$

$$w = \frac{1}{2}(-B(0)\xi n_3 e^{\frac{N_2^2 t}{Re}+\xi(n_3x+m_3y+l_3z)} - C(0)m_3 e^{\frac{N_3^2 t}{Re}+(n_3x+m_3y+l_3z)}).$$

Where $N_3^2 = n_3^2 + m_3^2 + l_3^2$, $N_1 = \mu N_3$, $N_2 = \xi N_3$.

Calculating the unknown p according to (5) we come to result

$$p = p_0 - gz, \qquad (40)$$

where g is the acceleration due to gravity.

The formulas (39), (40) represent a set of exact solutions of $3D$ Navier - Stokes equations, which determine the possible movement in a deep water under action of gravity. These formulas contain eight arbitrary constants $n_3, m_3, l_3, \mu, \xi, A(0), B(0), C(0)$. The first five of them are nonzero.

3.2. Solution 2

The second solution corresponds to the case of propagation of three plane waves when the collinearity condition for all three wave vectors is violated. In order to find the solutions in this case we require to study in more detail system of algebraic equations (31-36). Since the number of unknowns exceed the number of equations one can initially assume the feasibility of some additional relations that simplify further analysis. We propose the following

$$n_1 + n_2 = 0, \qquad (41)$$

$$m_2 + m_3 = 0. \qquad (42)$$

Based on assumption (42), equation (34) simplifies considerably. On the left-hand side only two nonzero terms remain. As a result, this equation take the form

$$l_3(n_2 + n_3)(n_2(l_2 + l_3) - l_2(n_2 + n_3)) = 0.$$

Due to (41) from the latter it follows

$$n_3 = -n_1 \frac{l_3}{l_2}. \qquad (43)$$

Taking into account assumption (42) equation (33) is also simplified and take the form

$$l_3(l_2 + l_3)(-n_2(l_2 + l_3) + l_2(n_2 + n_3)) = 0.$$

Due to (41) and (43), the last equation is certainly fulfilled.

Further we consider equation (31). Under condition of (41) it simplified and take the form

$$(-n_1 l_2(m_1 + m_2) - m_1 n_2(l_1 + l_2))(m_1 + m_2) = 0.$$

For most interesting case when $m_1 + m_2 \neq 0$, due to (41) we arrive to

$$m_1 = m_2 \frac{l_2}{l_1}. \tag{44}$$

Next we consider equation (32). Under condition of (41) it reduce to

$$n_1 n_2(-(m_1 + m_2)^2 + (l_1 + l_2)^2) = 0.$$

For most interesting cases when $n_1 n_2 \neq 0$, and $l_1 + l_2 \neq 0$ due to (44) we arrive to $l_1 = \pm m_2$. First we consider in more detail the case

$$l_1 = -m_2. \tag{45}$$

Due to (41-42),(43),(44),(45) the equations (31-34) are satisfied and equations (35-36) can be converted to the form

$$n_1^2(\frac{\theta^3}{\lambda} + \frac{\theta^2}{\lambda} + \frac{\theta}{\lambda} - 1 - \theta\lambda - \theta) - m_2^2 \lambda(1+\lambda)(1+\theta\lambda) = 0, \tag{46}$$

$$n_1 m_2^3 (2\theta(\theta-1)(\lambda+1) - \theta(\lambda-\theta)(\theta-1) - (\lambda-\theta)(1+\lambda) +$$
$$(1+\lambda)^2 - (\theta-1)^2) = 0, \tag{47}$$

where the new notation is introduced

$$\theta = \frac{l_3}{m_2}, \quad \lambda = \frac{l_2}{m_2}. \tag{48}$$

Equation (47) subject to $n_1 m_2 \neq 0$ leads to

$$\lambda = -\theta. \tag{49}$$

Taking into account the latter, it follows from equation (46) that

$$m_2 = \pm \frac{n_1}{1-\theta}\sqrt{\frac{2}{\theta}}, \tag{50}$$

where n_1 and θ are an arbitrary constants under conditions as $n_1 \neq 0$, $\theta > 0$, $\theta \neq 1$.

Thus, you can get a set of solutions to algebraic equations (31-36) depending on two arbitrary constants n_1 and θ. We will confine ourselves to considering only one case. Let's suppose $n_1 = 2$ and $\theta = 2$. Due to (35-45) and (50) we find the desired values of unknown wave numbers

$$n_1 = 2, \ m_1 = -4, \ l_1 = 2, \quad n_2 = -2, \ m_2 = -2, \ l_2 = 4,$$

$$l_3 = 2, \ m_3 = 2, \ l_3 = -4. \tag{51}$$

As a result, we arrive at the values of the main unknowns as

$$u = \frac{1}{2} e^{\frac{24t}{Re}} \left(-4A(0) e^{2x-4y+2z} + 4B(0) e^{-2x-2y+4z} \right),$$

$$v = \frac{1}{2} e^{\frac{24t}{Re}} \left(-2A(0) e^{2x-4y+2z} - 4C(0) e^{2x+2y-4z} \right),$$

$$w = \frac{1}{2} e^{\frac{24t}{Re}} \left(2B(0) e^{-2x-2y+4z} - 2C(0) e^{2x+2y-4z} \right). \tag{52}$$

$$p - p_0 = -gz + e^{\frac{48t}{Re}} \left(3A(0)C(0) e^{4x-2y-2z} - A(0)B(0) e^{-6y+6z} \right).$$

As a result, we obtain the set of new exact solutions of 3D Navier — Stokes equations. It contain three arbitrarily chosen constants $A(0), B(0), C(0)$.

3.3. Solution 3

Let us construct one more new solution, instead of (45) setting

$$l_1 = m_2. \tag{53}$$

At the same time, we leave in force the ratios (41-44).

On the case under consideration equations (31-34) are identically satisfied and (35-36) are transformed to the form

$$n_1^2 2(1-\theta) + m_2^2 \theta (1+\theta)^2 (1-\theta) = 0, \tag{54}$$

$$n_1 m_2^3 (2\theta(\theta+1)(\lambda-1) + \theta(\lambda-\theta)(\theta+1) - (\lambda-\theta)(\lambda-1)+$$

$$(\lambda - 1)^2 - (\theta + 1)^2)) = 0, \tag{55}$$

in which the designations (48) are retained.

From equation (55) we obtain $\lambda = -\theta$. The latter is similar to relation (49) in the case of solution 2. Further, based on (54), we arrive to

$$m_2 = \pm \frac{n_1}{1+\theta} \sqrt{-\frac{2}{\theta}}, \tag{56}$$

where n_1 and θ are an arbitrary constants under conditions as $n_1 \neq 0$, $\theta < 0$, $\theta \neq -1$.

As in the case of solution 2, it is possible to obtain a set of wave vectors containing two arbitrarily chosen nonzero constants n_1 and θ. We will confine ourselves to one particular case. Let's suppose $n_1 = 1$ and $\theta = -\frac{1}{8}$. A calculation by formulas (35-44),(53),(56) yields to

$$n_1 = 1, \ m_1 = \frac{4}{7}, \ l_1 = \frac{32}{7}, \ n_2 = -1, \ m_2 = \frac{32}{7}, \ l_2 = \frac{4}{7},$$

$$l_3 = 1, \ m_3 = -\frac{32}{7}, \ l_3 = -\frac{4}{7}. \tag{57}$$

As a result, we arrive at the values of main unknowns as the next

$$u = \frac{1}{2} e^{\frac{1089t}{49Re}} \left(\frac{4}{7} A(0) e^{x + \frac{4}{7}y + \frac{32}{7}z} + \frac{4}{7} B(0) e^{-x + \frac{32}{7}y + \frac{4}{7}z} \right),$$

$$v = \frac{1}{2} e^{\frac{1089t}{49Re}} \left(-A(0) e^{x + \frac{4}{7}y + \frac{32}{7}z} - \frac{4}{7} C(0) e^{x - \frac{32}{7}y - \frac{4}{7}z} \right),$$

$$w = \frac{1}{2} e^{\frac{1089t}{49Re}} \left(B(0) e^{-x + \frac{32}{7}y + \frac{4}{7}z} + \frac{32}{7} C(0) e^{x - \frac{32}{7}y - \frac{4}{7}z} \right). \tag{58}$$

$$p - p_0 = -gz + e^{\frac{2178t}{49Re}} \left(-\frac{72}{49} A(0) C(0) e^{2x - 4y + 4z} + \frac{1}{4} A(0) B(0) e^{\frac{36}{7}y + \frac{36}{7}z} \right).$$

Thus, another set of new exact solutions of 3D Navier — Stokes equations was obtained. It contains three arbitrarily chosen constants $A(0), B(0), C(0)$.

3.4. Free Surface Profile

The constructed above exact solutions corresponding to motion in deep water make it possible to obtain the equation of the free surface for each of the considered cases. For this, it is sufficient to put $p = p_0$ in the expression for pressure [1-2], where p_0 is an additive pressure constant. Further, setting $z = h$, where h represent the height above the horizon $z = 0$, we obtain an equation for the free surface profile.

Thus for solutions 1,2,3 in accordance we arrive to

$$h = 0, \qquad (59)$$

$$ghe^{2h} + A(0)(-3C(0)e^{4x-2y} + B(0)e^{-6y+8h})e^{\frac{48t}{Re}} = 0, \qquad (60)$$

$$ghe^{-4h} + A(0)(\frac{72}{49}C(0)e^{2x-4y} - \frac{1}{4}B(0)e^{\frac{36}{7}y+\frac{8}{7}h})e^{\frac{2178t}{49Re}} = 0. \qquad (61)$$

For solution 1 the free surface does not change with time and not depend on coordinates x, y. It represents a plane that coincides with the horizon plane.

For solutions 2 and 3, the situation is more complicated. The shape of the free surface is governed by implicit equation containing x, y, t value of h and other quantities. The value of h calculated by equations (60) and (61) depends on the coordinates x, y time t, and on the values of the parameters $Re, A(0), B(0), C(0)$. Calculations show that, depending on the parameters, the values ??of $h(x, y, t)$ can be either positive or negative.

4. DISCUSSION

Wave formation on the free surface of the water affects the dynamics of ship movement, the conditions of trouble-free navigation, service life of the metal structures of ship.

Often in works devoted to the formation and propagation of gravitational waves on the water surface only external observable factors such as wind strength, temperature, changes in atmospheric pressure are taken into account.

The research presented in this work show that the formation of waves is largely influenced by the movement of water masses at a depth.

From equations (60-61) it follow that the profile of the free surface is largely determined by the values of $A(0)$, $B(0)$, $C(0)$, which correspond to the conditions at a depth. In particular, pay attention to the following feature. Due to the fact that the parameter Re is positive, the time decrement of the exponents in equalities (60-61) is also positive and these quantities over time take large values. It is possible to choose such values of $A(0)$, $B(0)$, $C(0)$ that $h(x,y,t)$ becomes very large in absolute value, staying positive or negative. In this case, the considered motions on the free surface represent a model for formation of a rogue waves [12-14]. According to equations (60-61) such waves can be formed both in the form of a hump ($h(x,y,t) > 0$) or in the form of a depression ($h(x,y,t) < 0$). But in any case their lifetime is limited, since after a short period of time they are destroyed.

CONCLUSION

Study of the movement of water masses at a depth based on the 3D Navier - Stokes equations for isothermal incompressible medium was carried out. As a result various exact solutions are obtained and their properties were considered. Some regularities of formation and evolution of waves on the free surface are investigated. In particular, a model for formation of rogue waves was proposed.

REFERENCES

[1] Kochin N.E., Kibel I.A., Rose N.V., Theoretical fluid mechanics, Part 1, *Fismatlit, Moscow, Russia*, 1963, (in Russian).

[2] Loitsaynskiy L.G., Mechanics of Fluid and Gas, *Nauka, Moscow, Russia*, 1987, (in Russian).

[3] Lodizhenskaya O.A., Sixth Millennium Problem: Navier - Stokes Equations: Existence and Smoothness, *Russian Math. Serveys*, 58:2, 2003, 251-286.

[4] Fefferman Charls L., Existence and Smoothness of the Navier — Stokes Equation, *Preprint, Princeton Univ., Math. Dept., Princeton, NJ, USA*, 2000, pp. 1-5.

[5] Polynin A.D., Zaitsev V.F., Handbook of Nonlinear Partial Differential Equations (second adition). – *CRS Press, London, UK*, (2012).

[6] Aristov S.N., Knyzev D.V., Polynin A.D., Tochnye reshenia uravneniy Navie — Stoksa s lineynoy zavisimost'u ot dvukh prostranstvennykh peremennykh. – *Theoretical foundations of chemical technology.*, **43(5)** (2009), 547-566. (in Russian).

[7] Aristov S.N., Polynin A.D., Tochnye reshenia trehmernyh nestatsionarnyh uravneniy Navie — Stoksa. – *Dokl. Phys.*, **54** (2009), 316-321. DOI: 10.1134/S102335809070039, (in Russian).

[8] Koptev A.V., Integrals of Motion of an Incompressible Medium Flow. From Classic to Modern. – Handbook on Navier — Stokes Equations. *Theory and Applied Analysis*, Nova Science Publishers, New York, (2017), 443-459.

[9] Koptev A.V., Method for solving the Navier — Stokes and Euler equations of motion for incompressible media, *J. Mathematical Sciences.*, Springer, NY, USA, **2020**, 250(1), P. 254-265.

[10] Koptev A.V., Generator of solution of 2D Navier — Stokes equations, *J. Siberian Federal Univ., Math. and Phys.*, **2014**, 7(3), P. 324-330.

[11] Koptev A.V., Exact solutions of 3D Navier — Stokes equations, *J. Siberian Federal Univ., Math. and Phys.*, **2020**, 13(3), P. 306-313.

[12] Koptev A.V., Opisanie dvizhenia voln-ubiyts na osnove 3D uravneniy Navier – Stoksa [Description of the motion of rogue waves based on the 3D Navir - Stokes equations], *Vestnik gosudarstvennogo universiteta morskogo i rechnogo flota imeni admirala S.O. Makarova*, (**2020**), T. 12, 2(60), 279-288. (in Russian).

[13] Shurgalina E.G., Mekhanizm obrazovaniya voln-ubiyts v resultate vzaimodeystvia solitonov vnutrennikh voln v stratifitsirovannom vodoeme [The mechanism of formation of rogue waves in the result of

the interaction of solitons of internal waves in a stratified reservoir], *Isvestiya RAS, Mekhanika zhidkosti i gaza*, (**2018**), 1, 61-67. (in Russian).

[14] Shelkovnikov N.G., Volny-ubiytsy v okeane [Killer waves in the ocean], *Isvestiya RAS, Seriya Fisicheskaya*, (**2014**), T. 78, 12, 1621-1625. (in Russian).

INDEX

A

activation energy, 31
amplitude, 48, 91, 93
atmospheric pressure, 98
atoms, 6, 10, 11

B

base, ix, 61, 79
black hole, viii, 2, 13, 15
boundary value problem, 82
bounds, 8
burn, viii, 25, 28, 47

C

chaos, 14, 18
chemical, 26, 27, 31, 38, 58, 100
chemical reactions, 26, 38, 58
combustion, 26, 57
computation, 34, 58
computational fluid dynamics, 26, 58
conductivity, 28, 30
configuration, 48, 57, 68
convergence, 30, 40
correlation(s), 8, 11, 12, 15, 19, 20
cosmos, 12
critical value, 34, 43
crystal structure, 15

D

dark energy, 18
dark matter, viii, 2, 13, 15, 16
derivatives, 34, 65, 67, 68, 69, 70, 71, 72, 73, 74, 75, 87, 88, 89
differential equations, 68, 81
diffusion, 34
diffusion process, 34
displacement, viii, 7, 26, 27, 28, 40
dissociation, 35, 38, 54, 55, 56
distribution, vii, viii, 2, 8, 9, 12, 14, 15, 17, 18, 25, 27, 28, 31, 37, 38, 39, 41, 45, 46, 48, 49, 51, 52, 54
dualism, vii, 1, 17

E

electric charge, 5, 19
elementary particle, 3, 15

energy, vii, viii, 1, 4, 8, 9, 10, 11, 12, 13, 14, 15, 16, 17, 18, 20, 25, 26, 27, 28, 30, 31, 32, 33, 34, 35, 37, 38, 39, 40, 41, 42, 43, 44, 45, 46, 47, 48, 54, 56, 57, 58, 59, 60
energy conservation, 38, 39
energy density, 9, 14, 20
energy input, 27, 48, 57
energy supply, viii, 25, 26, 27, 28, 30, 31, 32, 34, 38, 39, 40, 41, 42, 43, 44, 45, 46, 47, 48, 56, 57, 58, 59
equality, 41, 64, 69, 70, 71, 72, 73, 74, 75, 76
equilibrium, viii, 12, 15, 19, 25, 27, 35, 38, 58, 59
Euclidean space, 11

F

fluctuations, 2, 6, 45
fluid, vii, ix, 32, 62, 63, 64, 79, 80, 81, 83, 84, 85, 99
force, 7, 16, 78, 84, 96
formation, 6, 27, 35, 50, 84, 98, 99, 100
formula, 37, 39, 40, 62

G

galaxy(ies), 19, 22
general relativity, 2, 5, 8, 13, 15, 18, 21
gravitation, 4, 5, 9, 14, 17
gravitational field, 12, 13, 17
gravitational pull, 7
gravity, ix, 5, 12, 14, 19, 23, 83, 84, 94

H

heat release, vii, viii, 25, 28, 46
heat transfer, 16, 59
human, viii, 2, 3, 6, 7, 8, 9, 11, 12, 20

human perception, viii, 2, 8, 9

I

ideal, viii, 5, 28, 33, 35, 36, 38, 40, 49, 50, 54, 55, 56, 61, 62, 63, 74, 76, 77, 78, 79, 81, 82
idealism, 4, 5, 6, 20
immortality, vii, 1, 3, 19
inertia, 5, 19
integration, 65, 67, 81, 86
intensity values, 39
ionization, 35, 38, 54, 55, 56

L

localization, 7, 45
low temperatures, 34

M

mass, vii, viii, 1, 5, 7, 8, 9, 11, 12, 14, 15, 16, 18, 19, 20, 29, 31, 35, 37, 40, 45, 57, 84
material surface, 6
mathematics, 5, 11, 15, 85
matter, vii, 1, 2, 3, 4, 5, 6, 7, 8, 9, 11, 12, 13, 15, 17, 18, 19, 20, 22
measurements, viii, 2, 10, 12
media, 82, 100
medicine, viii, 2, 10
metric spaces, 2
mixing, 15, 26
models, 9, 10, 13, 32, 33, 35, 38, 40, 56, 58
molecular weight, 35
molecules, 6, 27, 34, 35, 36, 54, 55, 56

N

natural science(s), viii, 2, 3

neglect, ix, 83
Newtonian gravity, 4
Newtonian physics, 5
nitrogen, 35, 54, 55, 56
nitrogen compounds, 35
nonequilibrium, 20
nonlocality, 2, 4, 11, 12, 15

O

objective reality, 3
ordinary differential equations, 91
overlap, 7, 11
oxygen, 35, 54, 55, 56

P

palliative, 10, 14
partial differential equations, vii, viii, ix, 64, 65, 83
photons, 19
physical characteristics, 62
physical features, 34
physics, vii, viii, ix, 2, 3, 8, 11, 12, 13, 15, 17, 18, 19, 30, 31, 61, 63, 64, 83
plasma current, 16
Plato, 3, 6, 15
population, 3, 10
probe, 5, 9, 12
propagation, 93, 94, 98

Q

quasars, 19
question mark, 64

R

reaction mechanism, 16
reality, 2, 4, 6, 8, 9, 11, 13, 14, 15, 19, 20

recognition, 10, 13
researchers, 3, 8, 9, 15, 20
resolution, 85, 87
restrictions, 68, 91
Riemann problem, 33, 38, 59
root, ix, 61, 64, 65, 68, 69, 74, 76, 77, 78, 80

S

science, vii, 1, 3, 7, 12, 20
self-knowledge, viii, 2, 6, 10
self-organization, 9, 10, 11, 14, 15, 16, 17, 18
shock, viii, 25, 26, 27, 28, 32, 34, 36, 41, 42, 45, 48, 50, 54, 56, 57, 58
shock waves, 32, 45, 57
simulation(s), viii, 26, 27, 28, 40, 58
solitons, 101
solution, 15, 30, 31, 32, 33, 34, 38, 42, 43, 45, 46, 47, 49, 50, 59, 63, 80, 81, 82, 87, 92, 93, 94, 96, 97, 98, 100
space-time, 10, 13, 18
specific heat, 29, 30, 35, 37, 39, 53, 54, 55
spiral galaxy, 15
standard model, 2, 4, 6
state(s), 8, 11, 28, 31, 32, 35, 36, 46, 47, 49, 50, 59, 63, 65, 71, 72, 73, 74, 76, 77, 78
structure, viii, 2, 12, 25, 27, 28, 32, 41, 68, 87, 90
superimposition, 6, 7, 11, 12

T

techniques, 34, 59
temperature, viii, 25, 26, 27, 29, 30, 32, 35, 37, 38, 42, 44, 45, 46, 47, 48, 49, 50, 51, 52, 53, 54, 55, 56, 57, 98
textbooks, vii, 1, 4, 6, 10, 14, 19
thermal energy, 16, 56

thermodynamic parameters, 37
thermodynamic properties, 35
tornadoes, 16
total energy, 29, 31, 36
transformation(s), 14, 31, 89
tribology, 63

U

uniform, 16, 40, 41, 46, 48, 59
universes, 5

V

variables, 29, 30, 31, 33, 34, 35, 36, 38, 62, 63, 64, 67, 68, 85

vector, ix, 14, 16, 17, 29, 30, 33, 34, 38, 62, 63, 76, 83, 84
velocity, ix, 18, 29, 31, 32, 33, 62, 63, 67, 68, 69, 76, 83, 84, 86
viscosity, ix, 28, 30, 32, 62, 83

W

water, vii, ix, 83, 84, 93, 94, 98, 99
wave number, 90, 91, 92, 93, 96
wave vector, 93, 94, 97
worldview, 3, 13

Y

yin-yang, 8, 11, 14, 15